Labyrinths

The Art of Interactive Writing & Design

CONTENT DEVELOPMENT FOR NEW MEDIA

DOMENIC STANSBERRY

Multimedia Studies Program
San Francisco State University

INTEGRATED
MEDIA
GROUP

An Imprint of Wadsworth Publishing Company
I(T)P® An International Thomson Publishing Company

Belmont, CA • Albany, NY • Bonn • Boston • Cincinnati • Detroit • Johannesburg • London
Madrid • Melbourne •Mexico City • New York • Paris • San Francisco
Singapore • Tokyo • Toronto • Washington

New Media Publisher: Kathy Shields
Assistant Editor: Shannon McArdle
Marketing Manager: David Garrison
Production: Hal Lockwood, Penmarin Books
Design: Image House
Composition: Myrna Vladic

Copy Editor: Alan Titche
Print Buyer: Karen Hunt
Permissions Editor: Peggy Meehan
Cover Design: Andrew Ogus
Printer: Malloy Lithographing, Inc.

Printed in the United States of America
1 2 3 4 5 6 7 8 9 10

For more information, contact Wadsworth Publishing Company, 10 Davis Drive, Belmont, CA 94002,
or electronically at http://www.thomson.com/wadsworth.html

International Thomson Publishing Europe
Berkshire House 168-173
High Holborn
London, WC1V7AA, England

Thomas Nelson Australia
102 Dodds Street
South Melbourne 3205
Victoria, Australia

Nelson Canada
1120 Birchmount Road
Scarborough, Ontario
Canada M1K 5G4

International Thomson Publishing GmbH
Königswinterer Strasse 418
53227 Bonn, Germany

International Thompson Editores
Campos Eliseos 385, Piso 7
Col. Polanco
11560 México D.F. México

International Thomson Publishing Asia
221 Henderson Road
#05-10 Henderson Building
Singapore 0315

International Thomson Publishing Japan
Hirakawacho Kyowa Building, 3F
2-2-1 Hirakawacho
Chiyoda-ku, Tokyo 102, Japan

International Thomson Publishing Southern Africa
Building 18, Constantia Park
240 Old Pretoria Road
Halfway House, 1685 South Africa

Library of Congress Cataloging-In-Publication Data

Stansberry, Domenic
      Labyrinths: the art of interactive writing & design: content
development for new media/Domenic Stansberry.
          p.   cm.
      Includes bibliographical references (p.) and index.
      ISBN 0-534-51948-2
      1. Interactive multimedia—Authorship.   I. Title.
QA76.76.I59S7 1997
808'.066006—dc21   97-19342

changing the way the world learns[SM]

To get extra value from this book for no additional cost, go to:

http://www.thomson.com/rcenters/img/img.html

thomson.com is the World Wide Web site for Wadsworth/ITP
and is your direct source to dozens of on-line resources.
thomson.com helps you find out about supplements,
experiment with demonstration software, search for a job,
and send e-mail to many of our authors. You can even
preview new publications and exciting new technologies.

thomson.com: *It's where you'll find us in the future.*

In Memory of Robert Bell

Founder and Director of the Multimedia Studies Program

San Francisco State University

CONTENTS

CHAPTER 1

Multimedia: Notes Toward an Aesthetic 1

Multimedia Defined 2
The Origins of Multimedia 3
    The Futurist Vision 3
    Revolution in the Sciences 8
The Nature of Multimedia 10
The Writer in Multimedia 11
Summary 12
Exercises 13
Endnotes 14

CHAPTER 2

The Design Process in Interactive Media: An Overview 15

The Writer's Role in Multimedia Design 15
Stages in the Design Process 17
    Stage One: Concept Development 18
    Stage Two: Script Design 23
    Stage Three: Production 27
The Design Team 28
Summary 29
Exercises 30
Endnotes 31

CHAPTER 3

The Design Shell 32

The Interplay of Form and Content 32
Content Design 34
    Design Metaphors 34
    Classification Schemes 40
    Flowcharts 44
Interface Design 46
Design in the Real World 48
Summary 49
Exercises 50
Endnotes 51

CHAPTER
4

Interactivity 52

Elements of Interactive Structure 54
   Screens 54
   Movement Between Screens 55
Patterns of Interactive Structure 56
   Linear Patterns 57
   Hierarchical Patterns 58
   Web Patterns 61
Adding Complexity to Interactive Structure 63
   Depth vs. Breadth 64
   Obligatory Cuts 64
   Multiple and Parallel Branches 65
   Stepladders and Glossaries 66
   A Final Word on Flowcharts 66
Summary 69
Exercises 69
Endnotes 70

CHAPTER
5

Arranging Content 71

Narrative Structure: Telling a Story 72
   Plot and Character 72
   Point of View 76
   Setting, Style, and Theme 77
Expository Structure: Conveying Information 79
   Introductory Elements 79
   Background Material 81
   The Main Body 82
   Concluding Elements 84
Summary 85
Exercises 86
Endnotes 87

CHAPTER
6

Scripting and Designing Computer Games 88

The Nature of Games 89
   Game Skills 90
   Elements of Games 90
Types of Computer Games 92
   Arcade Games 92
   Strategy Games 94
   Adventure Games 96
   On-line Gaming 98
The Game-Scripting Process 100

Developing a Game Concept   100
Creating a Concept Treatment   102
Composing the Full Treatment   107
Writing the Script   107
Handling Nonsequential Items   108
Summary   110
Exercises   111
Endnotes   111

CHAPTER
7

Educational Multimedia: Scripting from an Instructional Design Model   112

The Instructional Design Process   113
The Proposal Stage   113
Content and Structural Development   117
Script Design: Creating a Production Blueprint   126
Implementation and Revision   128
Models for On-line Education   129
A Final Look at Educational Multimedia   132
Summary   133
Exercises   133
Endnotes   134

CHAPTER
8

Scripting for Business Programs   135

Advertising Programs   136
Informational Programs   140
Training Programs   145
Format Considerations for Multimedia Business Programs   148
The Design Process for Business Programs   150
Summary   154
Exercises   154
Endnotes   155

CHAPTER
9

Technical Considerations in Multimedia Projects   156

The Relation of Conceptual Skills to Technical Issues   156
Platforms and Formats   157
Current Technical Limitations   159
Working Within the Technical Limits   162
Identifying the Proportions of Media Elements   163
Understanding Format Capabilities   163
Conserving Computer Memory   164
Understanding Software Capabilities   165

Expanding the Technical Limits   165
Software Tools for Multimedia Writers   166
Summary   168
Exercises   169
Endnotes   169

CHAPTER
10

Interactive Media and the Future   170

Dreams and Nightmares: The Ambiguity of the Future   170
    The Dark Side of Futurism   171
    Directions for Multimedia   173
Technical Advances in Interactive Media   174
The Multimedia Marketplace in the Future   176
The Future of Interactive Education   177
The Direction of Society in an Interactive World   179
Summary   181
Exercises   183
Endnotes   183

APPENDIX
A

Script Excerpts from the Interactive Cinema In the 1st Degree   *184*

Excerpts from the TREATMENT: Story Overview   184
Excerpts from the CHARACTER BIBLE   185
Excerpts from the PRODUCTION SCRIPT   187
Excerpts from the DIALOGUE SCRIPT   188

APPENDIX
B

Excerpts from the Production Script of the Children's Educational Program The Little Red Hen   *192*

Part I: From the OPENING SEQUENCE   195
Part II: From the PRE-READING ACTIVITIES   195
Part III: From the INTERACTIVE READ   197
Part IV: From the COMPREHENSION EXERCISES   198
Part V: From the CLOSING ACTIVITIES   200

APPENDIX
C

Web Site Specifications for the Apple Personalized Internet Launcher   *202*

I: Project Description   202
II: Content Areas   203
III: User Experience Elements   207

APPENDIX
D

Excerpts from the Treatment of the Adventure Game Dreggs *211*

The Environment 211
The Main Characters 213
Interactivity 216
The Introduction of Play 217

Glossary 219

Selected Bibliography 223
    Books and Periodicals 223
    CD-ROMs 224
    Web Sites 228

Index 231

About the Author 241

PREFACE

As the title of this book suggests, the focus of these pages is on the subject of writing and design for interactive media. This book emphasizes that part of the design process that comes first—before a single image has been scanned, or a frame of video has been digitized, or a programming sequence has been codified and entered. All of these activities are important, extremely so, and fascinating subject matter if described by the right person in the right forum. My own background, though, is as a writer and interactive designer. My concern is with content and how that content might best be shaped and arranged before the formal production process begins.

This book, therefore, is concerned with conceptual development. It focuses on the rhetorical and interactive patterns that bring content to life in this new medium. More specifically, it deals with the scripting process. It discusses the writing and charting techniques used to develop the proposals, treatments, production scripts, and design documents that act as the templates for interactive media. My hope is that these techniques will be useful not only to students but to professional writers, directors, producers, interactive designers, programmers, and all others who find themselves struggling to bring meaningful content to this new medium.

Two premises that are interrelated and often repeated throughout these pages underlie this book. The first premise is that new media are really not quite so new as they are made out to be. Rather, they are outgrowths of traditional media and utilize many familiar rhetorical techniques. Even the notion of interactivity itself, seemingly so new and startling, is new only in a relative sense, for it has its roots in certain perceptual changes that have been in progress in the arts and sciences since the early part of the twentieth century.

The other major premise underlying this book is the notion that certain fundamental principles of conceptual development will remain constant even as the technology continues to evolve. Compression rates, software capabilities, and delivery media may all change, but certain age-old fundamental principles will continue to guide the writing and design processes at their deepest levels.

For this reason, the main body of this book is concerned with identifying these fundamental principles and discussing the ramifications of their use. Since first beginning this project, it has been my goal to write a book that is useful not only as a practical guide, but that also placed the art of writing for multimedia in a historical, aesthetic, and intellectual context.

The first half of this book discusses and analyzes the design process, whereas the second half focuses on the use of that process in specific contexts. The text includes chapters on conceptual design and interactivity, and on the scripting process in entertainment, business, and educational contexts. There is as well a chapter on technical matters and a collection of sample scripts in the appendices. In the final chapter I discuss the future of this new medium as well as possible social implications of its growing use, whether for good or evil. My hope is that this will spur readers to think about how they work in this field, and about the consequences of their actions as writers and designers.

In a book of this nature, written over the course of several years, the writer inevitably owes thanks to a great many people from whom he has borrowed ideas or received advice. Good writers, it is commonly said, are not so much original thinkers as good thieves. Although I have avoided outright thievery here, there is a great deal of synthesis from many sources, all of which are listed either in the endnotes or in the bibliography, as appropriate. I would also like to take this opportunity to thank others who have helped with this project, either by providing certain opportunities or by commenting directly on the manuscript.

Particular among these is the late Robert Bell, the founder and original director of the Multimedia Studies Program at San Francisco State University. He gave me the opportunity to teach Writing for Multimedia in his fledgling program, and he also encouraged me to pursue the subject of aesthetics in this new discipline. He was a man of great personal vision, and his presence in the program is sorely missed.

I would also like to thank the poet Gillian Conoley of Sonoma State University. Gillian introduced me to the work of the Futurists and discussed endlessly with me the relationships among postmodern art, classical aesthetics, and interactive media. Similarly, I must express my gratitude to Peter Adair and Haney Armstrong, the filmmakers who gave me a chance to work hands-on within the industry, and from whom I learned a great deal about the relationship between interactivity and narrative form. I would also like to thank my editor, Kathy Shields, as well as her assistants, Tamara Huggins and Shannon McArdle. In addition, I owe special debts to Dr. Scott Fredrickson, University of Nebraska at Kearney; Merton E. Thompson, St. Cloud State University; Stephen W. Harmon, Georgia State University; Douglas B. Smith, California Polytechnic State University—San Luis Obispo; Greg Sherman, Emporia State University—all of whom spent a great deal of time and care reading and commenting on this work in its manuscript form.

Finally, I must also thank my students at the Multimedia Studies Program at San Francisco State University. They were the first to suggest that I write this book, and they have been my most constant source of knowledge, intelligence, and inspiration.

To all these people, as well as to other writers and designers whose work I have mentioned or cited in this work, I express my gratitude. They deserve much of the credit for whatever is good and useful in this book. On the other hand, as to any inaccuracies or confusions that might remain, whether they be the result of misstatements, clumsy thinking, or simple foolheadedness, the author takes full responsibility.

DOMENIC STANSBERRY

1

Multimedia: Notes Toward an Aesthetic

Collage is the art form of the twentieth century.
　　　　　—Donald Barthelme

The new life of iron and the machine, the roar of motorcars, the brilliance of electric lights, the growling of propellers, have awakened the soul, which was suffocated in the catacombs of old reason and has emerged at the intersection of the paths of heaven and earth.
　　　　　—Kasimir Malevich, 1916

Literature, art, drama—all have rich aesthetic and formal traditions born millennia ago, deep in human history. The same is true of the natural sciences, and of the various branches of human philosophy as well. Multimedia—if we consider it on its own, as a new field—is very much the infant. The field, it seems, is too young to have its own intellectual history, so those of us seeking to endeavor in it find ourselves alone, without a trail to follow—a frightening feeling. Or exhilarating, depending upon one's point of view.

The truth, however, is that new ideas are rarely as new as they seem. They come from somewhere; there is a history of other ideas behind them.

When looking for guiding principles, writers and interactive designers in multimedia can always turn to the other, older disciplines from which this new field emerged. Not all the older principles are relevant, of course, and sometimes the relevance is all but obscured by the context in which the principles are applied. The search for guiding principles is complicated by the fact that there are many who want to disclaim any relationship with the past whatsoever. This is new territory, they say (or perhaps the voice comes from within each of us), so the old rules do not apply. Thus the human desire to discover something new and startling comes into tension with another human desire: to link the future to the past.

As multimedia is a discipline shaped from many others—from computer science to cognitive psychology to visual arts to literature—the search for guiding principles is likely to be dynamic, elusive, slippery. The debate over what is useful from the past—which ideas, from which disciplines—strikes very much at the heart of the creative process. For that reason, it is a debate likely to continue for some time, with answers that change from time to time, circumstance to circumstance, even project to project.

This ongoing debate—this dialogue among disciplines, between the future and the past—is what makes the field so exciting, and so vexing. To understand the dynamics involved, it is useful to look at the recent history of the arts and sciences and how they come together in the presence of the computer to form this new field. Before taking this backward glance, however, we need first to formulate a definition of the word itself.

Multimedia Defined

The word *multimedia* reflects something of both the cultural milieu from which the discipline has emerged and its mixed allegiances. In the old audio/visual departments of educational institutions, and in corporations too, the adjective *multimedia* was used to describe lecture demonstrations that incorporated slide shows, blackboard presentations, and audio tapes. In the 1960s the word was used to describe psychedelic light shows that accompanied rock-and-roll acts in the Haight Ashbury district of San Francisco. It has since been used to describe installation art in museums, particularly futuristic exhibits that combine video and other media in a three-dimensional space.

It was not until the early 1980s, when artists and scientists started experimenting with computers as a way of manipulating light and sound, that the term began to be associated with computers. More recently, with the popularization of the term by the computer industry, it has been used in a broad, categorical fashion to describe almost everything that involves sight or sound in a computerized environment. At its simplest level, in its current context, then, multimedia involves the combined use of audio, visual material, and text in a computerized environment.

This definition, although good enough in many ways, is still incomplete. It misses an important fact recognized by the early developers of the personal computer: *The computer is not simply a tool but a medium.* What is unique and revolutionary about the computer as a medium is the addition of *interactivity*. Interactivity gives users the ability to choose their own path through the information—to navigate—and thus the ability to structure their own experiences. For the writer and interactive designer, this means learning how to structure material in a nonlinear fashion, so that it can be experienced from a variety of angles, no matter from which direction the user should approach.

The key element, then, that differentiates the computer from other media, is interactivity. Not only does multimedia involve the mixing of word, sound, and image in a computerized environment, it also gives the user the ability to navigate through and change—to interact with—that environment.

The Origins of Multimedia

There is a tendency to view multimedia as a field that has emerged from technology alone, as a child of Silicon Valley and the Digital Revolution. When the history of the field is discussed at all, it is discussed in terms of the development of the personal computer, with particular focus on the group of computer scientists associated with Douglas Engelbart in the Augmentation Research Center at Stanford University during the 1960s and 1970s.[1] It was these men, the story goes, who unleashed the magic from the box. Because of them, the computer is in the hands of the masses, a vehicle to catapult human consciousness into the twenty-first century. And because of the computer, society is on the verge of a major paradigm shift, a new way of thinking and of perceiving the world.

There is some truth to this line of thinking. It would be foolish to deny the possibilities for transformation inherent in the propagation of a device as powerful as the computer. It might also be argued, however, that we are not at the beginning of a major paradigm shift, but approaching the end, that the personal computer and the possibilities it has unleashed lie at the crest of a century-long shift in consciousness that has only now begun to reach fruition.

Some literary critics, such as Marjorie Perloff, have written about this shift, and trace its artistic corollaries to the early years of the twentieth century and to a tiny, scattered group of artists known as the Futurists. The center of this group, if such a center existed, was probably in Paris, though undercurrents of this movement were present in London, Moscow, Venice, and all the intellectual hubs of Europe. Among the artists moving in the vortex of the Futurist sensibility were the French poet and painter Blaise Cendrars, the poet Guillaume Apollinaire, and the American writer Ezra Pound. The man who gave the movement its name, however, was the Italian poet Filippo Tommaso Marinetti, author of a fiery, iconoclastic broadside on aesthetics called *The Futurist Manifesto*, published in 1909.

THE FUTURIST VISION

The Futurists viewed European civilization as a civilization weighed down by the traditions of the past—by an aristocratic class system, by the church, and by arcane national structures and obsolete systems of government. Art, too, they felt, had grown stale, had become a prisoner of an aged classicism bound by formalities and stifling notions of aesthetics. None of it, they thought, had the energy of the new gadgets—the automobiles and the radios—suddenly so prominent in the cities.

In *The Futurist Manifesto*, Marinetti called for an art that made a radical break with the past.[2] He wanted artists to lead society away from the old institutions and into the future. Artists should refuse to honor old boundaries, he said; they should juxtapose things that had not been juxtaposed before. Combine and then recombine. Artists should be rude and insistent, disrespectful of boundaries, just like the technology the Futurists admired.

Along these lines, in his various manifestos, Marinetti laid out two aesthetic notions: simultaneity and multilinear lyricism. *Simultaneity* refers to the notion, engendered by technology, that people, objects, and ideas can transcend physical, temporal, and social barriers to exist side-by-side in patterns that are somehow both fleeting and permanent. *Multilinear lyricism*, on the other hand, is the term Marinetti used to refer to the aesthetic sensation induced by the observation, creation, or manipulation of those simultaneous patterns. Both concepts were rooted in observations of perceptual changes brought about by technology, particularly in regard to how time and space were experienced on a daily basis.

Particular among technologies that influenced Futurist thinking were those having to do with communications and transportation. The radio and the telegraph, it seemed, made it possible for the human voice to be many places at once. The expression of ideas, conversation, and other forms of human communication were no longer limited by physical location, but suddenly transcended space and time. This simultaneity of being fascinated the Futurists, and they saw it happening all about them—in the newspaper photographs that carried images around the world; in the electricity that blurred the borders of night and day; in the trains, airplanes, and automobiles that hurled the physical self across the landscape. In other words, there was no longer one self, but many, a multitude of selves moving through a multitude of social and cultural universes. Reality had many strands that were diverse and experienced all at once. The feeling this simultaneity engendered is what Marinetti called multilinear lyricism.

As their symbol, the Futurists adopted the Eiffel Tower. With its iron struts, its fractal geometry, and its radio beacon at the top, it seemed the perfect fusion of technology and art (even though its builder had only functionality in mind: as an efficient, durable structure for the beaming of radio waves). The Futurists blurred the distinction between conscious art and unconscious expressions created by fortuitous juxtapositions they observed throughout the culture. They admired such juxtapositions in store windows, in the developing skylines, in advertising posters. Taking such juxtaposition as inspiration, Futurist artists worked across media—combining high culture with low, scrawling the printed word upon the formal canvas, examining one object from many angles, and placing common objects in museums. They ventured across traditional boundaries to create works of artistic expression that were based on juxtaposition, as opposed to classical symmetries. In other words, they explored notions of simultaneity and multilinear lyricism in their own art. They tried to capture the new state of consciousness engendered by technology and then present it back to the world around them. Paintings and collages by artists such as Umberto Boccioni and John Heartfield demonstrate the use of juxtaposition for such purposes (see Figures 1.1 and 1.2). In a similar vein, Francis Picabia's *Parade Amoureuse* (Figure 1.3) demonstrates the Futurists' fascination with machines.

Many artists associated with the Futurists later became part of the dadaist and surrealist movements, both of which shared many of the same sensibilities, particularly an aesthetic fondness for radical juxtaposition. Much of the influential art of the twentieth century grew out of these movements, including the spiraling figures of Miro, the cubism of Picasso, the collages of Joseph Cornell, the abstract expression of Pollack, and the installation art of Ed Keenholtz and Vito Acconci. Even the stark, repeated images of Andy Warhol—the Campbell's soup can, the faces of celebrity, the stamped-out images of advertising and industrial

Figure 1.1
Umberto Boccioni, *The City Rises*, 1910–1911. Oil on canvas. 6'6½" x 9'10½." In this painting, human forms seem to take flight and merge as the scaffolding of an industrial landscape rises in the background. Notice the juxtaposition of animal, human, and mythic forms, as well as the strong sense of motion, together suggesting both a liberation of the spirit and an undercurrent of violence. Paintings such as this demonstrate the Futurist notion that artistic, spiritual, and political liberation would accompany the new urban technology. By blurring the distinction between forms, Boccioni also suggests the dissolution of normal physical boundaries, thus illustrating a simultaneity of being, a union of spirit and matter that encompasses man, beast, and urban landscape. (The Museum of Modern Art, New York. Mrs. Simon Guggenheim Fund. Photograph © 1997 The Museum of Modern Art, New York, NY)

technology presented back to the culture as works of art—sprang from the same perceptual ground.

The sensibility engendered by the Futurists and their successors, more than once labeled the primary artistic mode of the twentieth century, has been prevalent in literature as well. Much of modern fiction, for example, has been obsessed with examining the multiple layering of consciousness using alternating points of view, pastiche, and collage-like juxtaposition. The obsession began with Virginia Woolf and James Joyce in the 1920s, perhaps, and has manifested itself in America in writers as diverse in their social concerns as John Dos Passos, William Faulkner, and Gertrude Stein. In the 1970s and 1980s the avant-garde in contemporary fiction became obsessed with exploding traditional ideas of narrative. The Futurist influence on literature in the late twentieth century continues with the

Figure 1.2
John Heartfield,
Cover for *Der Dada*
#3 (Berlin), 1920.
This collage juxtaposes
word and image,
placing common
household objects in
close relationship to
the human form, as
well as to automobile
tires and other objects
associated with
emerging industrial
technology.
(Photograph © 1996,
The Art Institute of
Chicago. All rights
reserved. © 1997
Artists Rights Society
(ARS), New York/VG
Bild Kunst, Bonn)

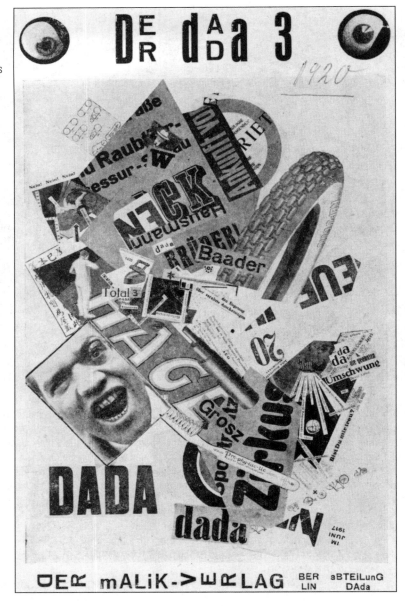

international popularity of writers such as Jorge Borges and Julio Cortazar, whose labyrinthine stories insisted on the existence of multitudinous parallel realities. Meanwhile, visual artists such as Peter Weibel (see Figure 1.4) and Nam June Paik had begun to explore similar themes in installation pieces, using television cameras and recurring loops of video to integrate the audience into the work of art.[3]

Figure 1.3
Francis Picabia, *Parade Amoureuse* (Amorous Parade), 1917. The Futurist love of machines influenced a number of artists, among them surrealists such as Picabia, who took the idea of the machine as subject matter. (Francis G. Mayer/Corbis)

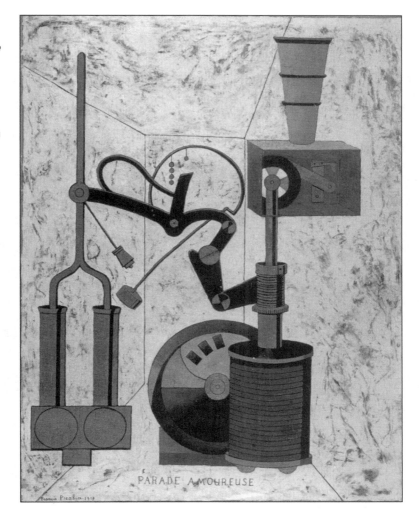

Many of the concerns of twentieth-century art—over point of view, over juxtaposition, over the angle of perception—are more than concerns in multimedia; they are inherent to it and as such move to the forefront. This is true because of the very nature of the way in which information must be presented in computer-driven environments. Multilinear arrangement of material is inherent to presentation of material in an interactive format. It is as if the personal computer is a manifestation of the early Futurist vision.[4]

To attribute computers directly to the Futurists, however, would be a mistake —or at least an oversimplification. Rather, the computer is an invention of a society intent on investigating certain forms of perception. Just as the arts have been focused on exploding classical structures, the sciences have been engaged in a full-scale attack on Newtonian physics and its ordered, deterministic view of the mathematical principles that underlie the structure of the universe.

Figure 1.4
Peter Weibel,
*Observation of
Observation:
Uncertainty,* 1973.
In installation art such
as this, which involves
three monitors, three
cameras, and a floor
mark, the audience has
an active role, changing
the artwork by their
presence and via inter-
action with the cam-
eras. Later, conceptual
artists such as Myron
Krueger added the com-
puter to these earlier
multimedia displays,
creating more interplay
between audience and
environment. Though
produced many decades
later, these installation
pieces are a direct out-
growth of the early
Futurist sensibility.
(Photograph © Peter
Weibel)

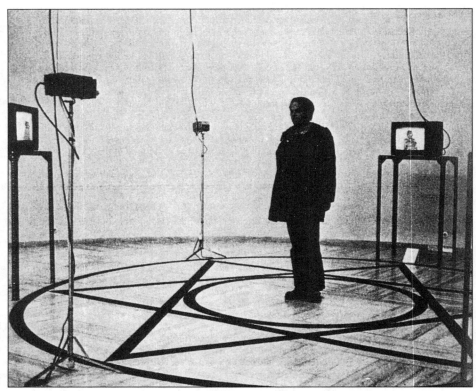

REVOLUTION IN THE SCIENCES

As science writer James Gleick and others have pointed out, the attack on Newtonian physics took place on three fronts. First, Einstein's theory of relativity
challenged the Newtonian illusion of absolute space and time. Then, quantum
theory challenged the Newtonian dream of a controllable measurement process.
More recently, chaos theory has undermined ideas concerning system analysis,
whether those systems involve weather formation, wildlife populations, or the
movements of automobiles on a freeway. Patterns once thought to be quantifiable and predictable have been proven otherwise, and randomness seems to be
present in almost every system.[5]

Mathematicians and physicists have discovered that a majority of natural systems—whether they involve turbulence patterns, human population growth, or
global warming—cannot be represented by linear equations, but instead are nonlinear patterns that oscillate over time. They may repeat themselves in a general
way, but not always and not precisely, and sometimes not at all. The old saying
"Numbers do not lie" has proved to be a dubious assertion.

According to Benoit Mandelbrot, a mathematician who popularized the concept of the fractal, even the length of a country's coastline is a subject of mathematical uncertainty, relative to point of view. "The notion that a numerical result

should depend on the relation of object to observer is in the spirit of physics in this century, and is even an exemplary illustration of it."[6]

In other words, like the artists and writers descendent from the Futurists, contemporary scientists and mathematicians have been exploring the relative nature of point of view and the possibilities of multiple descriptions of reality. To speak in the terms of Marinetti, they have been exploring the concepts of multilinear lyricism and simultaneity.

The computer sciences, too, have mirrored the intellectual movement away from classical structures. In the early days, a theorist named Pierre La Place fantasized about building a computer based on a Newtonian model of the universe. "Such an intelligence would embrace in the same formula the movements of the greatest bodies of the universe and those of the lightest atom: for it, nothing would be uncertain and the future, as the past, would be present to its eyes."[7] Many of the early pioneers in computer science, including John Von Neuman at MIT, were inspired by this Newtonian vision. The perfect modeling systems they hoped to create, however, were foiled by the complexity of the systems they were emulating—as well as by the difficulty in measuring them.

As early as the 1940s, computer scientists such as Vannevar Bush were moving away from the notion of building a computing machine that mimicked the Newtonian universe. Rather, Bush saw the computer as a machine that augmented human intelligence. In the past, machines had been regarded as a physical extension of the body, in the way that a garden shovel can be seen as an extension of the hand. Bush's notion was that the computer could be a mental extension of the human mind, an augmentation device.[8]

Bush's ideas influenced a generation of computer scientists, including Douglas Engelbart, who was a young Navy technician when he happened on Bush's 1945 *Atlantic Monthly* article, "As We May Think."[9]

Years later Engelbart told writer Howard Rheingold about his reaction to these new ideas:

> When I first heard about computers, I understood from my radar experience that if these machines can show you information on printouts, they could show that information on a screen. When I saw the connection between a television-like screen, and an information processor, and a medium for representing symbols to a person, it all tumbled together. I went home and sketched a system in which computers would draw symbols on the screen and I could steer through different information spaces with knobs and levers and look at words and data and graphics in different ways. I imagined ways you would expand it to a theater-like environment where you could sit with colleagues and exchange information on many levels simultaneously.[10]

It was Engelbart's notion to juxtapose elements from different technologies, linking the television screen with the computer. When, decades later, he had a chance to implement his ideas at the Augmentation Research Center at Stanford, he added a keyboard to the system and helped create word processing. The mouse pointing device, hypertext, and the concept of text "windows" all came out of Engelbart's ARC lab, as did the notion of the computer as a space through which the user navigated. In that space, the user might find not only text, but graphics, video, and sound as well.

Another computer scientist greatly influenced by Bush, and by Engelbart too, was Alan Kay. In his work at the Palo Alto Research Center in the 1970s, Kay helped develop the Dynabook, widely regarded now as the first personal computer. (Earlier computers had been room-sized devices that required special programming language to operate.) Kay, like Engelbart, wanted to put the computer into the hands of the everyday worker. Toward this goal, he pioneered the study of interface design, with particular focus on the use of graphics. Many of his ideas, particularly about how the computer might be used as a thinking aid, he gleaned from Jean Piaget and Jerome Bruner, educational psychologists who wrote about the importance of visualization and interaction in education.

Like others working in the media at the time, Kay was also greatly influenced by the writings of Marshall McLuhan:

> Though much of what McLuhan wrote was obscure and arguable, the sum total was a shock that reverberates even now. The computer is a medium! I had always thought of it as a tool, perhaps a vehicle—a much weaker conception. What McLuhan was saying is that if the personal computer is truly a new medium, then the very use of it would actually change the thought patterns of an entire civilization.... What kind of thinker would you become if you grew up with an active simulator connected, not just to one point of view, but to all the points of view of the ages represented so they could be dynamically tried out and compared?[11]

Kay's observations sound as if they could as easily be descendent from Marinetti as from Vannevar Bush. The same is true of another visionary thinker in the computer sciences, Theodor Nelson, who postulated the imaginary realm of Xanadu—a vast world of interconnected information, a three-dimensional library of thoughts and images that contained all the world's cultural history.[12] Though Nelson's Xanadu never came to pass, one can see its real-life descendant beginning to take shape today in the World Wide Web and other computerized information systems.

The Nature of Multimedia

It is tempting at this point to tie the discussion together in some neat and satisfying way—to say that the computer is an outgrowth of the Futurist sensibility, a culmination of a vision first articulated by Marinetti in 1909. There might be some truth in such a conclusion, but again it would be overly simple.

On a day-to-day basis, artists and technologists have, throughout this century, most often worked in isolation from one another. But this does not mean that they were not, in some deeper way, in touch with some shared cultural sources. The phonograph, the telephone, the television set—all of these ubiquitous devices have changed the consciousness of the world; the simple act of flipping the remote control of a television takes the user on a trip through multiple narratives. Such experiences are commonplace, are part of the spirit of the times, the international zeitgeist. Neither scientist nor artist can take credit or blame. Rather than being the product of a technology in isolation—or the discovery of a particular individual—the idea of a multitudinous reality is something that has emerged out of the culture itself. It is part of the consciousness of the future we have been busily imagining for the past century.

Even so, multimedia, as disciplines go, is young, barely into its infancy. If the history of ideas is any indicator, multimedia will no doubt transform itself several times before anything resembling a standard set of principles emerges. Such a transformation seems particularly likely in a field in which the technology has yet to fully mature and in which technology so directly affects the creative process.

In the meantime, some interesting questions present themselves. What is the role of the writer and conceptual person in this new medium? What is interactivity, and how does it affect traditional notions of expository and narrative structure? How do we conceive three-dimensional spaces for the exploration of information, and how much control, ultimately, do we give the user in exploring those spaces? And finally—and probably most importantly—what kind of world are we creating as we take this new medium into our households, our businesses, and our artistic and cultural institutions? Is it ultimately the kind of world we want to live in?

The Writer in Multimedia

Such questions are vast in their implications; they are easy to raise but much more difficult to answer. For writers and interactive designers, a starting point may be to remember that the computer is more than a tool. It is a medium, and as such it carries its own intrinsic message. How that message is shaped, its connotations and denotations, are inevitably bound to the medium itself. In this case, the medium carries with it the element of interactivity, the aura of something that has been crafted by users themselves. Though random exploration and multiplicity of experience seem to be characteristics of the medium, part of the designer's job may be to find the center point, to lead the user through the maze. In other words, the artist's job here may be what it has always been—rather like the magician's—structuring the illusion, bringing order out of chaos.

If the thrust of the literary avant-garde over the past century has been to explode classical traditions, one of the most exploded has been the Aristotelian notion of narrative. In the 1960s and 1970s narrative was reexamined by fiction writers, some of whom declared the novel dead even as they invigorated the form with explorations into the use of point of view, multilinear narrative, and collage-like techniques. These techniques seem to have particular relevance today, as people explore similar notions in multimedia. Despite the attack upon it, narrative survived and remained an important element of writing, even in the work of writers who eschewed it. People still tell stories, and these stories still have beginnings, middles, and ends. The contributions of the Futurists aside, the fundamental classical structures still have usefulness, just as do the fundamental physical laws of Newton. Apples still fall from trees, after all, no matter Einstein's law of relativity, just as carpenters manage to build plumb walls and level houses using the simplest of tools.

The point of the Futurist sensibility ultimately is not that everything we know about the universe is wrong, but that there are different angles of perception.

Within this universe there exist narrative and associative patterns alike, sometimes standing on their own, sometimes overlapping. The same is true in multimedia, in which the structuring of information often involves the fusing of interactivity with traditional narrative and expository concerns.

In it broadest definition, the universe includes not only everything that *is*, but also everything that *was* and *will be*, and even that which *could be*. In such a vision the ancient exists alongside the yet undiscovered, not only in causal relationship, but also haphazardly, the result of infinite juxtaposition. In the same way, multimedia writers may find themselves crossing disciplinary lines, using the principles of narrative and exposition but also exploring the worlds of computer games, visual design, and computer programming. Ultimately, what emerges from this enterprise may be a whole new definition of what it means to be a writer. After all, writing a book is a far different experience than writing a script for a virtual world in which users create their own juxtapositions, their own order, even their own disorder.

Any study of aesthetics in any field involves not just looking at one idea of how things are made, but many. It involves looking at how notions of beauty and form and proportion change over time and how contradictory ideas can exist simultaneously—within a culture, even within a person or a work of art. Similarly, the study of aesthetics in multimedia must consider not only the pertinence of classical forms, but also a multitude of opposite possibilities, embracing chaos and order, beauty and ugliness, sense and nonsense, and all ramifications in between.

SUMMARY

In its current usage, the word *multimedia* is often used to describe any media program that employs audio, visual, and textual elements in a computerized environment. One of the key elements separating multimedia experiences from other media is interactivity, the ability of the audience to control and manipulate their own experience. As such, the computer is more than a tool, but rather a medium with its own unique characteristics, forms, and tendencies.

The aesthetic of this new medium has its roots not just in one discipline but in many and represents a fusion of art and technology. This fusion has its antecedents in the work of the early Futurists, in particular *The Futurist Manifesto*, published by Filippo Tommaso Marinetti in 1909. In that work, Marinetti posited the notions of simultaneity and multilinear lyricism, which lay the foundation for experiments in collage, radical juxtaposition, and point of view that dominate much of twentieth-century art. Related movements have pervaded the sciences and humanistic disciplines as well, undermining earlier Newtonian and deistic views of the universe and replacing them with a far more relativistic vision.

Inspired by the writings of Vannevar Bush in the new field of computer science after World War II, Douglas Engelbart began experimenting with the notion of developing a computer that in many ways fulfilled the Futurist vision of melding human and machine. These men saw the computer as an augmentation device that could increase the speed and capabilities of the human mind. Building on the ideas of Engelbart and Bush, Alan Kay helped create the Dynabook, the first personal computer. Kay was also influenced by the ideas of media theorist Marshall

McLuhan as well as educational thinkers such as Piaget and Bruner. The result is a new medium, interactive and cross-disciplinary in nature, that relies as much on the techniques of association and juxtaposition as it does on classical rhetorical patterns.

E X E R C I S E S

1. Think about your daily routine and list the names of any technological devices that induce the sense of simultaneity and multilinear reality described in this chapter. How many of these devices existed fifty years ago? A hundred? How have these devices changed the way people perceive the world?

2. Choose one device from the list you generated for Exercise 1, and then describe its purpose and how that purpose is accomplished. How does use of this device affect your perception of the world outside you? Of other people and places?

3. Consider again the list you generated for Exercise 1. Eliminate one of the items from the list, and consider how your perceptions of reality might be different if that device did not exist. Imagine a different device that accomplished the same task, and describe how it might cause you to perceive the world differently.

4. In the past few years, a great deal of attention has been paid to technological developments in interactive media and the sweeping changes they will bring to society. What earlier technological developments have promised similar types of transformations? In what ways are they similar? In what ways are they different?

5. This chapter maintains that the major intellectual and artistic currents of the past century were informed by the Futurist sensibility. What is that sensibility? In what ways did it manifest itself in the arts and sciences over the last hundred years? Are there other trends that run counter to those suggested in this chapter? If so, consider another way of interpreting the intellectual trends of the past century to show that Futurism is not so dominant a mode as suggested here.

6. Consider the key elements in the definition of multimedia. Which of these elements are also found in television? Which are not?

7. In what ways is the computer a tool? In what ways is it a medium?

8. Both Marshall McLuhan and Alan Kay have argued that the computer has the potential of altering consciousness, of changing the way in which we perceive reality. What things might computers help us see differently? What things might the computer hinder us from noticing at all?

9. Toward the end of this chapter the text raises some questions about the computer and the type of world its expanded use is likely to create. What hopeful possibilities do you see ahead? What dreadful ones? In what ways might society avoid the potential dangers you foresee?

ENDNOTES

1. Howard Rheingold, *Virtual Reality* (New York: Touchstone Books, 1991), 72–75, 81–83.
2. Marjorie Perloff, *The Futurist Moment* (Chicago: University of Chicago Press, 1986).
3. Rainer Fuchs, "Self Construction—The Staging Self-Reference," in *Self-Construction* (Vienna: Museum Moderner Kunst Stiftung Ludwig, 1995), 43–46.
4. Pamela McCorduck, "Sex, Lies and Avatars: Interview with Sherry Turkle," *Wired*, April 1996, 108–110.
5. James Gleick, *Chaos: Making a New Science* (New York: Viking, 1987), 5–6.
6. Gleick, *Chaos: Making a New Science*, 97.
7. Ibid., 14.
8. Vannevar Bush, "As We May Think," *Atlantic*, August 1945.
9. Rheingold, *Virtual Reality*, 68–85.
10. Ibid., 74.
11. Alan Kay, "User Interface: A Personal View," in Brenda Laurel, ed., *The Art of Human–Computer Interface Design* (Menlo Park, CA: Addison-Wesley, 1990), 191–207. See also Rheingold, *Virtual Reality*, 84–86.
12. Rheingold, *Virtual Reality*, 180.

2

The Design Process in Interactive Media: An Overview

This chapter provides an overview of the design process, with particular emphasis on its three major stages: concept development, script design, and production. Before we examine these stages in some detail, however, it is important to note that writers in interactive media have many functions, all of which tend to overlap with the work of others involved in the design process. For that reason, the writer's role requires some further discussion and definition.

The Writer's Role in Multimedia Design

To begin with, multimedia itself is a very elusive and variously used term, encompassing a wide variety of projects, from full-fledged theatrical presentations to simple interactive disks. As a consequence there is a great deal of variation in whether and how writers are used. Producers and project managers tend to use writers differently, depending on their own creative tendencies as well as the needs of a given project.

A second impediment toward an absolute definition of the writer's role is that the design process is often very fluid and collaborative, involving a great deal of trial and error, of starting over, throwing away, and then beginning again. As such, as with any creative process, it can be foolhardy to lay out steps and establish roles in too absolute a manner, the way you might for the assembly line manufacture of automobiles. (Even that process, though, has its own vagaries and unpredictable elements.)

A final difficulty rests in the fact that interactive design is a young field with a terminology that is not always consistent, involving the cross-disciplinary use of language.

Despite all this, the fact remains that writers have a vital role in the design process. That role may be variable as the wind, but it can also be vigorous, rewarding, and central to the shape of the project as a whole. There are also certain guideposts to help, including the stages of the design process itself, which are outlined and discussed in this chapter. Though the design process may not always follow a straight line, nonetheless certain consistencies, certain patterns of development, and certain elements are common to most projects, no matter how circular the route to completion. It is these constancies on which we will focus here.

On many interactive projects, the writing and the design process are inevitably intertwined, so much so that at times it is impossible to discern one activity from the other. In other words, content and the structure in which it is presented are inextricably bound. This may not seem like a particularly overwhelming observation, but it does have a profound effect on the way design teams are put together—both on personnel and the assignment of tasks.

To understand this, it might be useful to look at the roles of creative personnel in several fields. Traditionally, in software development—and in the computer industry at large—the primary creative roles have fallen to technical people, to programmers and engineers. The writer's job comes later, after the product has been created, and that job has been to provide documentation, to write user manuals, and to help market and sell products. In other words, the primary function of the writer in software development has not been to create something but to explain how it works.

This is in stark contrast to the entertainment and publishing industries, in which writers are usually at the heart of the creative process. This may be truer in books than in film—where the director, as *auteur*, controls the ultimate vision —but it is still true that a writer's work is almost always at the source of that vision as a touchstone and a guiding force. When software people come together with entertainment personnel on a multimedia project, it is easy to see how sensibilities might collide.

Neither the software nor entertainment model offers an accurate template for the role of the writer in interactive media. Rather, in the latter the writer becomes a member of a team whose task involves creating a script addressing a wide number of design concerns.

Design for multimedia is a very broad field, encompassing a spectrum of diverse skills. Included are graphic design, which has to do with the look of things on the screen; software design, which has to do with the coding and structure of the program; interface design, which has to do with the way in which the software program manifests its interactive features to the user; and interactive design, which has to do with the structure of information pathways down which the user navigates.[1]

Of these different elements, it is very often true that the sheerly visual and the sheerly technical are the things with which writers have to concern themselves the *least*. This is not to say that writers do not write animation sequences, nor that they should be ignorant of technical parameters. Rather, the writer's primary job is to sculpt the shape and structure of content, and as such the writer is most involved with the interactive design—the charting and flow of information pathways.

The central idea here is that the different design elements—the interface, the interactive pathways, the graphic look, the content, and the technical limitations—all affect one another and are intimately related. This is a collaborative medium, and thus the disciplines of writing, visual arts, and computer design tend to overlap in ways that are often messy. As the technology matures, these overlapping roles may eventually be codified and separated, in the way that the roles of producer, director, writer, cinematographer have all been codified by the Hollywood film industry. Perhaps this will happen in multimedia as well, though there are plenty of reasons to think differently. Much in modern society suggests in fact that the opposite will be true, that skills will continue to overlap. Much of this overlap, for better or worse, is due to the computer itself and the way in which it enables a single individual to work in many conceptual environments.

Stages in the Design Process

Most design processes can be broken down into stages that are very ancient, in the sense that they seem to be common to most creative endeavors. Multimedia is no exception.

The first stage—often referred to as *concept development* in contemporary diction, but known as "the art of invention" in classical terms—involves the task of inventing and evaluating ideas that might be suitable for presentation in the medium.

The second stage, known as *script design*, is an extension of the invention process in which the initial concept is explored and revised, culminating in a blueprint for the project. In multimedia, this blueprint typically consists of conventional script material, but might also contain flowcharts, storyboards, and software prototypes.

The third stage, *production*, involves the construction, assembly, and editing of the various elements specified in the blueprint. The production process itself can be broken down into steps or phases. In a multimedia project, which can involve the presentation of picture, sound, and text in a computerized environment, the assembly process often requires a wide range of artists—from software programmers to graphic artists to musicians and cinematographers.

These three stages may appear neat and discrete, but in truth they overlap a great deal.[2] As in any endeavor, people are sometimes forced to change their ideas or approach when confronted with the real-life terrain. Such changes are not indicative of failure, but comprise a necessary part of the design process.

Given this fact, there are plenty of good reasons for producers to keep writers and interactive designers involved throughout all phases of a project, following an idea from concept through the scripting phase and into production. Even so, this does not often happen. Part of the reason concerns another real-life exigency: money. It is expensive to pay conceptual people to stick around through the production phase. Another reason has to do with the roles, politics, and personalities of production teams, which consist of people who—like most of us—tend to be territorial about their specialties.

So, in practical terms, most of the writer's work in interactive media occurs in the first two stages of the process, involving the initial concept and the development of that concept into a full-fledged interactive script. Along the way, the writer/designer will meet often with other people, including content experts, graphics people, producers, and programmers.

STAGE ONE: CONCEPT DEVELOPMENT

The first stage in the design process, concept development, involves defining a concept, conducting research, and creating a preliminary interactive structure known as the design shell. Again, as with so much in interactive media, these activities are not discrete one from the other but instead part of a back-and-forth process.

Defining a Concept   Before a media project can begin, somebody has to have an idea they want to communicate and an audience they want to reach. This is as true for interactive media as it is for traditional media, no matter whether the project is crassly commercial or has some higher artistic or educational goal.

It sometimes happens that a writer is also a producer and in a position to shepherd his or her own ideas through the design process. More often, though, a writer is brought onto a project involving the ideas of others, as a person whose specialty is development, elaboration, and refinement of the initial concepts.

Regardless, the process is the same. So are the requirements. The concept must be focused in terms of both content and audience, and it must also be suitable to the medium. Along these lines, the producer must choose the delivery format most likely to reach the intended audience, whether that format be floppy disk, CD-ROM, the Internet (see Figure 2.1), or some new format like Digital Video Disk. Because different formats have different media capabilities, this decision greatly impacts the design process. The design team needs to be aware of the possibilities and limitations of their project, just as television and film producers must be aware of the corresponding constraints in their media.

Whatever the format, however, writers remain an integral part of many design teams. A computer game producer, for example, might call upon a writer to help develop a dramatic concept for a multimedia courtroom drama. Or a Fortune 500 company might decide that the best way to recruit new employees on college campuses is through an interactive brochure on floppy disk. Or an educational publisher might need help adapting children's books for multimedia presentation in the classroom.

All three of these projects, in the form they have just been described, represent ideas in an early stage of conceptual development. Each of them has a general subject matter and some inherent sense of audience. Even so, the content is not yet very specific. None of the ideas, as expressed here, is presented in a way that makes apparent why it should be published in an interactive medium.

In interactive media, a good concept will have at its heart a design idea that allows the user an avenue into the material, a unique angle of exploration.

If, for example, a game producer's idea of bringing a courtroom drama to CD-ROM is to be a viable one for interactive media, the user must have some angle into the material that isn't provided in other media. In the case of *In the 1st Degree*

Figure 2.1
Major Formats for
Interactive Media.
The choice of format is
a primary consideration
in interactive media.
Different formats have
different media
capabilities and thus
affect the writing
process. Formats are
discussed in greater
detail in Chapter 9.

FLOPPY DISK

The familiar 3.5-inch floppy is capable of handling a handful of full-screen color graphics, a sound bite or two, and maybe a simple animation. Its capabilities can be increased by sophisticated programming, and by spreading a single project across several disks. This is still the most widely available interactive format.

CD-ROM

This compact disk, read-only format, is similar in looks to the audio CDs used with home stereos. Each disk can store an hour or more of compressed video files, as well as numerous full-screen graphics, audio, and bookshelves full of text. The actual proportions depend upon compression schemes, the skill of the programmer, and the relative amounts of each media element. Video takes up the most space, so the less video used, the more room for everything else. Until recently, this was the most media-rich interactive format, though this has begun to change with the next generation of digital video disks and the continued development of the World Wide Web.

DVD (DIGITAL VIDEO DISK)

This is a new format, similar to CD-ROM, but capable of holding several times the amount of digitized video and animation. Hardware and software manufacturers are currently introducing this format into the market, including a DVD-ROM version intended for use with the computer. Although developers promise greatly enhanced video quality, the level of consumer interest at this point remains uncertain.

THE INTERNET

The thousands of individual Web sites, home computers, and host servers that together make up the World Wide Web hold perhaps the greatest potential for multimedia delivery. The host servers provide a virtually infinite storage ground for video and audio files. Even so, delivery of this medium has been hindered by the transmission capabilities of the telephone wires that serve as the major link between personal computers and remote servers. Developers are experimenting with a number of alternative wiring systems, software programs, and direct streaming techniques to circumvent these limitations. These techniques work best when the user's main modem connection takes place through alternative wiring systems, such as ISDN, T1, or T2 lines, rather than the conventional phone system.

—a multimedia drama built along this premise—the designers addressed this issue by having the user play a starring role: the prosecuting attorney.

What makes this an interesting starting point for an interactive medium is that, right off the bat, the role of the user is apparent. There is an interactive conceit built into the program from the very start. But that is only the beginning of concept development. The writer/designer then must ask a series of other questions. Does anybody really want to experience this? Once the user has entered the program, what shape will the interaction take? What is the information (or in this case, the storyline) that drives the program and makes it interesting? What other programs are similar to this?

Some of these questions will not be fully answered until a good deal further along in the process. Even so, before an idea can be taken seriously as a legitimate design concept, a great deal of serious grappling with ideas must take place.

Such grappling is not unique to interactive media. Before writing a book, writers research their subject, take notes, refine their ideas. Fiction writers do this, as do film writers. They become familiar with their medium, their genre, and their audience. They make lists of ideas. They throw some ideas out, return to others. They refine the ideas over and over. They make outlines, and then refine the ideas again. They write a few scenes. They may even start over.

The process of invention is time consuming. It involves tentative forays into the later stages of the creative process, then a return to earlier stages, and continual trial and error. The same is true in multimedia.

If one thing is unique about conceptual development for computer media, it is the interactive element. There is the content on one hand, and the user on the other. The meeting of these two in an interactive environment must be firmly envisioned if a solid concept is to result.

In this early stage of a project, one job the writer often has is to write a concept statement—to boil the essence of the concept down to a statement, a few sentences at most, that describes the project in such a way that the content, the genre, the audience, and the interactive conceit are all apparent.

This does not mean that the concept necessarily becomes frozen or that there is no more room for development or refinement. Constraint is not the intent of a content statement; rather, it is to provide a guidepost for the design team, so that they know where they are going and what it is they are trying to achieve. The more precise that statement, the better.

Producing a concept statement is a collaborative process that involves the design team's careful consideration of two issues: the content itself and the interactive structure in which the content will be experienced.

Conducting Research   Both the development and implementation of a concept require research, which can take several forms: research that investigates content, research that investigates structure, or research that investigates audience.

Research that investigates content explores the subject matter and the material that forms the basis of the program. This is true as much for games as it is for business and educational projects. In the example of the courtroom drama, the designers needed to research the trial process—to attend trials, take notes, go through court records, and interview lawyers. They even went so far as to retain a local DA as a consultant. In a perhaps more mundane example, a writer/ designer working on an employee-recruiting brochure would conduct research by meeting with the corporate client, reading the corporate literature, and interviewing the company's recruiting agent.

The writer's job as researcher is to learn as much as possible about the subject matter at hand. More than that, the writer must decide which information is important and which isn't; what to keep, what to throw away—all of this to achieve focus.

The same sort of thoroughness is required when it comes to research of structure. In this type of research, the writer/designer analyzes other projects that are similar in nature and have been produced with a similar goal in mind. For the previous example, the writer might ask, What other interactive brochures are out there? What do they have in common with paper brochures? How is the information structured and translated to an interactive medium? What is good about the competing work? What is not so good?

When working on a project, some people hesitate when it comes to examining similar projects. They fear, perhaps, that they will be too influenced, that their scope of vision will be limited by a glimpse of what has been done before them. They want to start with a clean slate. While such protectionist instincts are understandable, it might be wiser to look at what others have done, for there is a great deal that can be learned. Most artists of any merit, in whatever field, are constantly studying the work of their peers, learning and evaluating. The refusal to engage in this process more likely signals creative ossification than wild originality.

The third type of research, standard in larger software companies, is research into the audience. This includes marketing research as well as getting advance reactions of the people who will eventually use the program in their homes and offices.

Typically, in software development, products are user-tested in two stages. The first stage, called an *alpha test*, evaluates user reaction to an early prototype of the program. At a more advanced stage of development, audience reaction to a second (or beta) version of the software is evaluated in a *beta test*. In each stage, the reactions of the audience are quantified and incorporated into the design process.

Increasingly, though, software developers are feeling the need to bring the potential audience in at earlier and earlier stages, sometimes even before the design process begins.

Creating Interactive Structure: The Design Shell As previously mentioned, a hard look at the content is one of the first chores in the development of any program. Of equal importance is the creation of a *design shell*: the structural concept that serves as the basis for the interactive design (see Figure 2.2).

A key element of the design shell is known as the *interface*. We will deal with the concept of interface, and other aspects of shell design, more thoroughly in Chapter 3. For now, suffice it to say that the interface is the place where the user and the computer meet. Or, to say it another way, in physical terms the interface consists of the computer screen itself, plus the items that are placed on that screen for manipulation by the user.

Figure 2.2
Design Shell Triangle. The design shell is a conceptual tool for refining the design before scripting begins. It consists of three elements: content, interactive structure, and interface.

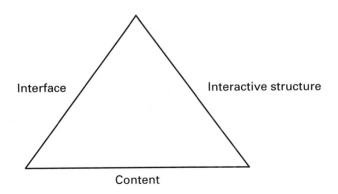

Perhaps the simplest and most common methodology for placing information on the screen relies upon the development of *classification schemes* (see Figure 2.3), in which information is broken up into categories that manifest themselves as visual areas on the screen. An interactive version of a university catalogue, for example, might be organized according to school (Arts & Letters, Humanities, Physical Sciences, etc.), and the opening screen would then present these category choices to the user on the interface.

Another notion often used in the creation of an interactive structure is the *design metaphor* (see Figure 2.4), which involves the creation of an imaginary or parallel environment, a place where information is explored by interaction with places, things, events, or characters (as opposed to unadorned classification schemes).[3] This is achieved primarily by comparing the information to something else to which it has similarities. The most ubiquitous example of a design metaphor, perhaps, is the Macintosh interface, which is organized along the metaphor of a desktop. In this conceptual world, the computer screen is likened to something familiar and ordinary: the top of a desk. Information is created on documents, stored in folders, and thrown away in the trash.

In interactive design, the conceptualization of the design shell is as important as the content itself. Though the intricacies of the interactive pathways may not be decided until further along in the process, true scripting cannot begin until the basic structure of the shell has been determined. This involves seriously addressing the following questions: Is there a solid content outline? What are the major elements of the design shell? How will the information appear on the interface?

Figure 2.3
An Example of a Classification Scheme. This Web site makes use of a simple classification scheme, in which one clicks on one of the listed topics (left side of screen) to learn more about David Byrne and his music. (*David Byrne Website*, http://www.bart.nl/~francey/th_db2.html; designed by Frank Veldkamp, Studio Zimbra.)

Figure 2.4
Design Metaphor.
This screen shot of
an oasis from *Myst,*
the popular adventure
game, demonstrates
the use of a design
metaphor. Rather than
navigating through
categories, players
discover clues while
roaming an imaginary
deserted island, using
the computer mouse
to move from place
to place, picking up
objects, and moving
back and forth in time.
(*Myst,* Broderbund.)

STAGE TWO: SCRIPT DESIGN

The second major stage of the design process concentrates on structuring, developing, and adapting the content in the context of an interactive structure. In general, this stage involves the creation of a sequence of draft documents that increase in specificity and clarity until there is sufficient detail to enable a production script to be written.

The earliest documents at this stage, which often focus exclusively on program content, are referred to as content treatments or content documents. Another type of document produced as part of script design focuses on the interactive pathways; such documents consist largely of flowcharts and written specifications elaborating the details of the design shell. All these early flowcharts and content treatments eventually form the basis for a third type of document, the production script, which includes all the details necessary for the creation of the program itself.

Content Treatments   Once a concept has been formulated, and the design team has some notion of the perimeters of the project, the design process has reached a point at which a treatment might be written.

Written *treatments*—early draft documents highlighting the main content, interface, and interactive features of a project in process—are common in many fields, though the term is probably most often used in reference to film projects. In Hollywood, a treatment is a straight prose rendition of the story, stripped of scripting format, that focuses on the story line and the characters. For writers, a treatment is a way of coming to grips with the material, in a short, condensed format, before struggling with the longer script format. It is a way of making sure all of the elements are in place. For producers and directors, it provides a way of working with a writer in process, as well as a means of deciding whether or not to develop a project at all. In the film industry, as in other industries, polished treatments are often used as part of the proposal process, as documents that help secure funding.

Treatments in multimedia are used similarly—as a means for developing the initial concept and getting a look at the primary elements that will shape the program. Treatments tend to be relatively short documents, though that is not always so.

In general, multimedia treatments concern themselves with some blend of the following elements: *program content, interactive structure, interface design, and media elements* (see Figure 2.5). There are, however, no given rules for the exact proportion in which these various elements will be discussed in a treatment; nor are there standard formatting conventions beyond those of individual producers and project directors. How a treatment looks, and what precisely it contains, instead depends on the needs of the project and the writer's common sense.

Many treatments, for example, are concerned with content alone. A treatment for an interactive encyclopedia of insects may contain nothing but the names of the critters and descriptions of their salient characteristics. (In this case, the document might be referred to as a "content document," not a treatment, but its purpose is similar.) In a complex multimedia drama, the treatment might consist of character sketches, a narrative presentation of the primary plot elements, a discussion of plot points, and other story considerations similar to the background writing often done for a television series.

Figure 2.5
Elements of
Treatments.

PROGRAM CONTENT
Concept statement
Program theme and purpose
Summary of major content elements

INTERACTIVE STRUCTURE
Classification scheme or design metaphor
Overview flowcharts

INTERFACE DESIGN
Descriptions or drawings of primary interface
Descriptions or drawing of primary navigational tools

MEDIA ELEMENTS
Visual look and feel
Balance of video, audio, and text elements
Technical and software considerations
Platform and format

A treatment can also contain some initial description of both the interactive structure and the interface. At this point in the design process, such remarks usually relate to the general structure, the breakdown of major categories, and the primary pathways. These written descriptions may be accompanied by flowcharts, sample screen shots, and other graphic material.

Finally, treatments may also touch upon media concerns, such as the look and feel of the project as well as the balance among video, animation, audio, and other media elements. Some treatments even contain information on programming specifications and other technical matters. In truth, though, the longer and more detailed a treatment becomes, the more it moves away from being a treatment and starts to become the script itself. This growth, from treatment into script, is a natural process, and the line between the two can often be very hazy indeed. Part of the process by which a treatment evolves into a production script involves exploring interactive pathways.

Exploring Interactive Pathways   For the most part, writers work in one of two modalities, narrative or exposition: They are either telling a story, or explaining how something works. There are, of course, shadings between these two modalities, and classical rhetoricians have developed an extensive terminology for the various techniques by which information can be arranged. Even so, for practical purposes, narrative and exposition are the primary structural modes. Both modalities tend to move forward in time—narrative by relating a sequence of events, exposition by tracing out the logical development of an idea.

Multimedia introduces a new element into traditional structures: the notion of interactivity. Interactivity offers a new role for the audience, and in so doing disrupts the natural thrust forward, the movement in time, from which argumentative and narrative presentation derive much of their power.

In traditional media, the goal is to persuade, to create fully the illusion of the fictional world, or to present ideas in ways such that they build one upon the other with undeniable force. The audience is swept away by the flow of ideas or by the movement of the drama on stage. Their intellect and emotions alike are affected; their mind is changed, they weep and take action. Meanwhile, the writer, to paraphrase James Joyce, stands backstage, aloof, paring his fingernails, playing the role of God.

In the eyes of some, interactive media challenges this basic role. Instead, they claim, it empowers the audience to take control of the material, and thus changes the fundamental relationship between the creator and the audience.

Before investigating this further, lets take a look at three different ways of looking at interactivity.

On the first and simplest level, interactivity allows users to react to an experience. A suggestion box, therefore, can be viewed as a simple interactive device. Applauding or hissing at a theatrical performance is another ancient form of interactivity, although plays and TV talk shows that incorporate the audience into the performance offer greater interactivity. (Whether this means higher quality is, of course, another issue.)

On the second level, interactivity provides a venue for the audience to choose their own path through the program, according to their needs or intuitions. A practical example is an automatic teller machine. Another, more maddening example, perhaps, is the type of automated answering services used by many companies, which direct callers through an apparently endless series of choices in search of contact with a real human.

A third way of looking at interactivity is that it provides the audience with a way to alter the program altogether, so that there is the potential for every user's experience to be different. The audience, in effect, becomes the author, and the material itself is ever changing. Here, one might think of William Gibson's *Agrippa*, a computer-based story that offers the reader several navigation options, and then destroys itself as it is read.

Though not all interactive programs evaporate in such avante-garde fashion, everyone working in interactive media is faced with the same basic consideration: how to give the user control of the information and still have it make sense. Even though interactivity may empower the user, it can also disrupt continuity and logic, not to mention narrative presentation. A key question, then, becomes how much is random and how much is controlled. The design team must decide what is essential for the user to experience and what is okay for them to miss.

Such decisions are at the heart of interactive design, and the writer is often the person who must grapple with them most frequently, because they are intimately related to the presentation of content. The process often involves not just writing, but also the creation of flowcharts and other ways of visually mapping pathways down which the user will travel.

What emerge, very quickly, are contradictions. The more options provided for the user, the less cohesive the structure and the more the user tends to get lost. On the other hand, the fewer the number of decision points, the more key information disappears. The decision to assert narrative, linear control, undermines the rationale for placing the project in an interactive medium.

Like much else in this chapter, the details of interactive design and flowcharting will be discussed later in the book (see Chapter 4). For now, it is perhaps best to make note that, as is the case with so much else in interactive media, the process of sketching out the interactive pathways is a fluid one, a dance between content and structure that has its own rhythms and demands. Although there may be some truth in the notion that interactivity changes the relationship between creator and audience, it does not change it completely. The user may roam, but in order for the experience to be meaningful, there must be a structure—and that structure has to be constructed in a very thoughtful and deliberate way. It may well be that interactivity does not give the user control so much as it gives the *illusion* of control—and the ancient puppet masters are still at work behind the scenes.

The Production Script   The complete melding of content with the interactive design takes place during the creation of the production script. At this point, the dialogue script, audio and visual cues, and flowcharts are put together into a document that will eventually act as a template for programmers, video personnel, graphic artists, and others. As with the treatment, the size and format of the full script tends to vary according to the nature of the project.

On a large, complex multimedia drama, the full script might be several hundred pages long and can consist of several documents. One of these might be a design document detailing the nature of the interface and the programming specifications. Another might be a dialogue or scene script containing dialogue segments numbered and coded to correspond to a flowchart. Added to this will be other flowcharts and diagrams that convey the overall structure, the individual paths, and the relationship between scenes. In addition, there may be storyboards and a functionality script, the purpose of which is to specify the navigational possibilities from any given screen. (Four scripts are provided in the appendices at the back of this text.)

On other, less lengthy projects, the full script might be much shorter. A single document with a screen-by-screen breakdown of functionality and content might suffice. There is a great deal of variability in how scripts are structured and how they look on the page, as well as in the terminology used to describe a particular script format. On one project, for example, the term *design document* might refer to the detailed production script; on another, the same term might be used to refer only to programming and interface specifications; on a third, the term might not be used at all. Regardless of the terminology, the unifying requirement for all scripts is that the material be clearly and consistently presented, so that the people who work from them can implement them well. Toward this end, almost any multimedia project of any size has on its staff a person whose job is to monitor the various codes and tags that function as identifying labels on interactive scripts.

STAGE THREE: PRODUCTION

Writers do not usually engage directly in the production process, as this part of the design process—which concerns the creation, assembly, and editing of digital material—is typically the domain of videographers, animators, programmers, and other specialized personnel. There are exceptions, though. Because of the nature of multimedia production, certain aspects of production will sometimes begin before scripting is finished. For example, backgrounds and other visuals may be rendered early on, and the writer may be asked to tailor the writing to those renderings. Similarly, a writer may consult with a programmer and be called upon to make changes in a script in light of technical considerations. On some projects the writer may be called upon to write on-screen text, headlines, and captions for a project already into production. Usually, however, the most direct link the writer has with the production process concerns prototype versions of the program created for the sake of user testing.

User Testing　It is rare in human endeavor that what we hope to create, what we plan and design, comes out exactly as we envisioned. Though frustrating at times, such discrepancies are often among the hidden rewards of creative effort. To see one's creation emerge from the design and take on a life of its own—and to see it used and interpreted in ways never anticipated—is one of the sweetest pleasures that can accompany the work of people in almost any field.

Even so, we do our best to anticipate what the final product will be, so that most of the surprises are pleasant ones and the creation serves its desired purpose. Toward this end, writers outline their material, architects build models, software engineers create beta versions, and marketing gurus conduct taste tests at the local shopping center. Despite all this planning, though, there comes a time when those closest to the creative process can no longer see their own project clearly. Then they need outsiders to take a clearer look. They need someone with a fresh point of view—if not someone objective, at least somebody different. Though design teams on interactive media can consist of personnel with varied expertise, this diversity is still no assurance that they are able to judge their project effectively. People working together on a project tend to become inculcated in its logic and blind to its failures.[4]

It can also be extremely difficult to judge the success of a multimedia project from its script. The flowcharts, the content documents, and the dialogue segments exist together in a web of cross-references, their sense and meaning eventually determined by spatial context as much as logic. An interactive script is, in essence, a one-dimensional rendering of a multidimensional object. To get a clearer view, it is necessary to build a model. Toward this end, a mock-up, or software prototype, is created.

This prototype can be very simple, containing only words and elementary graphics put together in a way that approximates the interface as it will eventually be experienced by users. Sometimes all that appears on the screen are the navigational icons accompanied by excerpts from the script itself. Because of this dependency on the written word, such prototypes are sometimes referred to as *text engines*.

Prototypes give the design team a sense of how the program is working. They add another dimension to the charts and functionality instructions present in the script. In addition, prototypes can be updated and revised with the script; they can also be modular in structure, so that they grow and evolve with the script itself. Another advantage of a prototype is that it can be passed along to people outside the design team. It can be given to its intended audience so that designers can judge and gauge users' reactions. It is at this point—hovering between script design and the commencement of serious production—that the writer is mostly likely to have a final chance to make revisions.

Eventually, of course, the project must pass from prototype to production. In some projects this is a bigger leap than in others, particularly when video and expensive animations are involved. In such instances, prototyping becomes even more important. Unless the writer is also a technical person or performs other roles in the production process, the writer's role tends to diminish during production. Some producers, though, like to keep a writer on call until the final wrap —to revise on-screen text, polish dialogue, and for other consultation purposes.

The Design Team

Just as the design process is a very fluid one—with one stage informing another, the boundaries hazy and overlapping—the same can be said about the division of labor. In interactive media, the lines between tasks are not always precise, nor are those among personnel. But even though the design process tends to be extremely collaborative in nature, certain roles still need to be performed. Although one person may perform more than one role or work in combination with others on a single task, nonetheless certain predictable work tasks need to be done.[5]

The team includes, of course, the producer, sometimes called a project manager, who is responsible for organizing and scheduling the project, managing money, and keeping the personnel together, amiable and focused. Then there is the writer, who is responsible for generating and arranging the content, and who does so in consultation with various media, programming, and content experts. As often as not, the writer's role is that of synthesizer: a person who listens to others and gathers together their ideas, who focuses and provides shape and structure. In some projects, the person who does this might not be called a writer at all, but rather an interactive designer or an information designer. In addition, most projects have graphic designers who supervise the visual look and feel, and frequently there is also call for audio and video technicians. Finally, of course, there is a programmer, whose job it is to make the whole thing happen.

Not all of these personnel will have equal weight on the design team. Some will participate only as consultants, and often several skills will be focused in one individual.

In the early days of computer gaming, it was often true that the programmer, working alone, was the primary creative force, handling not only the software design but the writing, graphics, and interactive structure. Increasingly, however, the conceptual work has become the domain of design teams consisting of writers and visual artists, with the programmers performing an implementation function. This is not always true, of course, nor should it be.

Design teams are staffed according to the project. In any given project, certain concerns might predominate. In one project it might be the visual image; in another it might be the delivery of a dramatic idea; in a third it might be the design of the software itself. This dominant concern will affect the roles of the personnel, so that writers working in multimedia have to be flexible—as willing to take on a supporting role as that of the creative lead.

S U M M A R Y

The writer in interactive media is most often responsible for the development and arrangement of content, particularly the breakdown of that content into interactive pathways. However, because interactive design often involves the overlap of graphic, software, and interface considerations, the role of the writer tends to vary from project to project and often overlaps with the roles of other people on the design team.

The design process itself is a reiterative process, starting with concept development, moving on to scripting and design, and culminating in production. During initial concept development, the task is to develop the idea for an interactive program in a way that considers not only content, but also interactive structure and how that structure will be manifested on the interface. Often, that concept is boiled down to a simple, concise statement that serves as a guidepost for the project. In the script design stage, the design shell is elaborated upon and expanded. Though the exact form of treatment documents tends to vary according to the needs of a particular project, they tend to address the same issues: content, interactive structure, and interface concerns. A treatment often addresses other media concerns as well, such as the look and feel of the project and the balance between audio and visual elements. Sometimes a treatment touches upon programming concerns as well, though these are more often addressed separately. The full script, or production script, delineates the graphic and media elements—realm by realm, screen by screen, activity by activity—including visual descriptions, dialogue rendering, text elements, navigational possibilities, and flowcharts. The production script, otherwise known as a design document, is what the programmer uses to put the project together.

Most projects also allow for user testing, in which the software is given to its potential audience in prototype form. At this point, the design team has a chance to evaluate how well the program is reaching its audience and whether or not it is achieving its desired effect. Prototyping and testing are vital parts of the design process and often take place throughout the project. They are yet another element in a collaborative process that requires the writer to be both a creative synthesizer and a source of original vision.

E X E R C I S E S

1. Define and discuss the following terms: interface, multimedia, interactivity, functionality.

2. Call up a local institution that relies on an interactive phone message system to handle its incoming calls. Call several times, working your way through the system, exploring the various navigational options until you feel you understand the system. When you are done, make a diagram, or flowchart, that illustrates the structure of the system. How effective do you think the system is? What are its advantages and disadvantages? How could it be structured differently?

3. A number of everyday devices—televisions, telephones, automatic teller machines—are in effect interactive devices, allowing users various degrees of control over information and services. Discuss the nature and degree of interactivity in each of these devices.

4. List the primary stages of the design process as outlined in this chapter. Describe what happens at each part of the stage.

5. List the four primary elements of a multimedia treatment. Describe each element in detail and discuss the relative importance of each.

6. Suppose someone asked you to design a multimedia program about your home town. (a) What research would you do? (b) Who would be your content expert? (c) What would be the focus of your content document? (d) What might be the interactive structure?

7. In order to be useful to the design team, a solid concept statement must deal with two primary considerations: content and structure. Given this fact, create a concept statement for the multimedia program about your home town.

8. Come up with an original idea for an interactive program. Test its viability by taking it through the process outlined in Exercise 6. When you are finished, write a solid concept statement.

9. If you were asked to put together a design team, what personnel would you include? Which of these personnel would play key creative roles, and which would be assigned implementation roles? Justify your decisions.

10. In preparation for a group exercise, come to class with an original idea for an interactive program. In class, break into groups and discuss the various ideas. Then evaluate each idea in terms of its suitability to the medium. Next, choose one of the ideas to develop more fully. At this point, each person in the group should take on a role in an imaginary design team. Working together, develop your idea, taking it through the various stages of the design process as outlined in this chapter until you come up with a workable concept statement.

11. Using the concept statement developed by the group in Exercise 10, write a brief (three to five pages) treatment.

E N D N O T E S

1. Brenda Laurel, ed., *The Art of Human–Computer Interface Design* (Menlo Park, CA: Addison-Wesley, 1990).
2. The process discussed here generally follows the film model and is based on personal observation and experience in the industry. It is similarly delineated in Ken Fromm and Nathan Shedroff, eds., *Demystifying Multimedia* (San Francisco: Vivid Publishing, 1993), 66.
3. S. Joy Mountford, "Tools and Techniques for Creative Design," in *The Art of Human–Computer Interface Design*, 25–31.
4. Howard Rheingold, "An Interview with Don Norman," in *The Art of Human–Computer Interface Design*, 5–10.
5. Michael Korolenko, *Writing for Multimedia* (Belmont, CA: Integrated Media Group, 1996), 104.

3

The Design Shell

The previous chapter provided an overview of the writing and design process, noting its various stages and elements. The purpose of this chapter is to elaborate on one part of the process: the conceptualization of the design shell.

As mentioned in Chapter 2, concept development for interactive media involves more than just content. It must also take into consideration interactive structure and the manifestation of that structure on the interface. During concept development, the task of the writer/designer is to blend these elements—content, interactivity, and interface—into a seamless entity known as the design shell.

This chapter elaborates on the primary techniques used to achieve this synthesis, including two conceptual tools useful in information design: the design metaphor and the classification scheme. Also covered here are some elementary flowcharting techniques that aid in the conceptualization of the interactive structure and some basic information on interface design. The chapter ends by discussing some of the constraints typically faced by designers and writers during project development.

Before turning to these matters, however, it is helpful to begin with some background information regarding a consideration vital to the conceptualization of the design shell: the interplay of form and content.

The Interplay of Form and Content

Multimedia writers and designers invariably find themselves enmeshed in the clash between form and content. It is a struggle that cuts across many disciplines, it seems, and is inherent in the creative process.

The novelist John Gardner, writing in *The Art of Fiction*, maintained that as soon as the first sentence in a novel was written, the writer had invariably placed the book inside a literary form, or genre. That genre, he argued, brings with it expectations, conventions, and limitations—in short, a history. Thus, the philosophical dimensions of a given novel—its themes, its content, and its emotional impact on the reader—are to a considerable degree determined by the formal requirements of the genre.[1]

Other writers have examined a similar notion from different angles, observing that each medium has certain innate qualities that shape the audience's experience of the intended message. This is true whether that medium consists of print, illustrative art, or moving images. A rendering of a news event on television, for example, may contain qualities of immediacy unavailable in print, but it lacks the reflective quality necessary to establish context and meaning. The differences in what is possible in different media are in fact so startling that it led to Marshall McLuhan's now-famous observation that the media is not simply a vehicle for a message, but rather that the medium *is* the message.[2]

Though from very different perspectives, McLuhan and Gardner are in fact addressing the same question. For artists and communicators, it is a question that lies at the heart of the struggle to give shape to our ideas: What comes first, form or content? Or, to restate the same notion another way: Does the idea inevitably determine the structure, or is it the other way around?

Like the age-old question concerning the chicken and the egg, this question too is probably unanswerable. It may be that the answer does not really matter. Rather, it is more important to observe that form and content are so inextricably intertwined as to be two ways of looking at the same phenomenon, examining it from different angles. To change the form is to change the content; the opposite holds true as well.

This does not mean that form and content are synonymous. Rather, they are two elements of the same phenomenon that are in constant interplay; one cannot exist without the other.

Content is the idea, message, or information that is to be communicated; by contrast, *form* is the structure into which that information is cast. The fusion of content and form determines the nature of the expression and how it is experienced.

Anyone who has worked on any project in just about any medium—whether it be a simple theme paper for a freshman composition class or a complex multimedia program—knows how the original idea is altered and shaped by its interaction with form. The tension is so familiar that most serious writers and artists regard it not only as inevitable but essential, an integral part of the process by which ideas are discovered, tested, and refined.

In traditional media, we have a pretty good idea of the formal possibilities. In journalism, the classic form encountered on nearly every front page is the inverted pyramid, which demands that the (relatively few) most important facts—the apex of the pyramid—come first in the article; these are followed by the (relatively many) smaller details—the wider base of the pyramid. This, of course, is not the only journalistic form. Editorials, daily columns, and feature articles each have their own rules and stylistic traditions that are followed assiduously. Each allows for different types of commentary and analysis.

The same is true in other disciplines, including literature, television, and film. Over the years various formats and structures have become standardized and familiar. It is also true that these forms undergo constant evolution; they are not static. Old formats are discarded and new ones created.

Multimedia does not have this rich formal history. Some may regard this as a blessing, because we are free to do as we please. Such freedom, though, carries a price, for without forms to guide us there is no shape; without shape there is no meaning. A simple glance at the vast majority of multimedia titles created over the last five years tells us that the absence of formal constraints has done little to produce work of more than transitory interest. (Moreover, that interest usually stems from the works' technical accomplishments, not their content.)

This observation may seem odd, especially when we consider how obsessed multimedia developers are with questions of form, of how to build and construct the software that drives their programs. Such formal questions, though, tend to be focused on the production process, not on the shape of the content. Perhaps this is because the medium is constantly redefining its technical possibilities, and in this environment it is very difficult to know the boundaries within which we are communicating. This will remain a problem until the technology matures.

Content Design

Even so, recognizing the great freedom arising out of the interplay of form and content does not mean that multimedia is without guidelines for the structuring of content. Some such forms we have inherited from older media, just as film inherited some of its dramatic conventions from the theater. Other forms are unique. Like film, multimedia is a visual and auditory medium. Unlike film, it is spatial, offering—through its interactive elements—an environmental space in which the audience can move, seemingly of its own volition. A name sometimes used for this space is the *design shell*, a term writers and designer use in the conceptual stage to refer to the computer environment they are trying to construct for users. As mentioned in Chapter 2, it can be useful to conceptualize the design shell in terms of a triangle composed of three elements: content, interactive structure, and interface. It is important to remember, however, that within the design shell these elements do not function as individual entities but as a unified whole.

Though guidelines for the construction of the design shell may not be as formalized in multimedia as they are in other media, nonetheless certain techniques are used with enough regularity to suggest that they are somehow intrinsic to the art of communication within this medium. One of these involves the use of design metaphors to conceptualize an interactive structure; the other involves the use of classification schemes. We consider design metaphors first.

DESIGN METAPHORS

When the human mind regards something for the first time, the impulse is to put that new thing in an old category—to find ways in which it resembles something else with which the mind is already familiar. One of the primary ways the mind makes such comparisons is through the use of metaphors.

Metaphors—words and expressions that define a thing by comparing it to something else—are as commonplace as dirt. Our everyday language is full of them. When we say wealthy people are "drowning in money," we are comparing their wealth to the vastness of a body of water. Although a Naked Lady, a colorful flower that grows on the roadside in California, is neither naked nor a lady, its name compares the bloom's beauty to that of the human form. And home gardeners refer to young plants that spring up without the gardener's having planted them as "volunteers."

Spoken language is not only full of metaphor, but is itself a metaphoric system using sound to indicate objects or ideas. Written language is a visual representation of that system, just as maps are metaphors for physical space. A metaphor is a way of recognizing something by comparing it to something else, or, to borrow from designer Thomas Erickson, it is "a verbal and semantic tool for conceptualizing both superficial and deep similarities between situations, people, places."[3]

Just as metaphor is used in language, it is also used in design, sometimes consciously but often not. In design contexts metaphors spring from our natural tendency to look at something in the light of similar things that already exist. Thus the first automobiles were thought of as horseless carriages, which explains why early autos retained many of the design features common to the horse-drawn carriages of the nineteenth century, even though some of those features were neither practical nor safe once an engine was attached.

Similarly, the early designers of multimedia have turned to books and cinema for metaphoric reference. Early multimedia works have borrowed heavily from those conventions. A common software design technique has been to create visual representations of familiar objects to be used and manipulated in a new context on the computer screen. Thus word processing programs employ the conventions and symbols of the typewriter, graphics programs use paint boxes, and flight simulators have screens designed to mimic the cockpit of an airplane.

The use of metaphor to create a design shell is also referred to as *design by symmetry*. In this technique, also described by Erickson, the designer searches for a similarity of structure or process and seeks to juxtapose concepts that are similar at a deep level. Once that underlying symmetry is established, the designer attempts to extend the symmetry further, using what is known about one area to suggest new ideas about another.

There are, of course, limitations to this technique. Metaphors cannot be extended infinitely, because the symmetries eventually break down. Such symmetries, at a certain point, can be irrelevant or misleading. A metaphor, after all, is not the thing itself, but only a representation of it.

The end goal of a design is not to create a perfect reflection of the thing being represented. Rather, the goal is to create an environment for users that is consistent, that has about it the air of familiarity, a place users can enter and intuitively feel their way around.

Spatial Metaphors   Perhaps the most common types of metaphors encountered in interactive media are *spatial metaphors*: those having to do with the arrangement of objects in a physical space. This approach originates in a way of thinking about the computer that goes back to Alan Kay and other computer pioneers who wanted to find ways of breaking down the cognitive wall between humans and computers.

Up until the mid-1980s, the need to compose typewritten commands in specialized computer languages kept the computer out of the reach of most users. As early as the 1940s, computer theorists had been speculating on ways to break those barriers down, to bring the computer into everyday life. One idea for doing this—seemingly out of a science fiction novel—involved melding the machine to the body through headsets and gloves that allowed a person to enter the world of the computer and to manipulate objects in that world.[4] One of the first practical applications of this notion was object-oriented programming, which allowed the user to manipulate objects on the computer screen instead of typing commands. This way of maneuvering has now become commonplace.

The tool that allows us, metaphorically, to move about on the interface nowadays is attached to almost every computer. That tool, the mouse, in effect functions as an extension of the human hand, breaking down the barriers between the user and the computer, allowing us to figuratively reach inside the screen and move objects around. Voice recognition and motion sensors will, in time, break down the cognitive wall even further.

For this reason—and perhaps because of the inherently schematic nature of software design—one of the favorite metaphors for multimedia designers has

Figure 3.1
A Spatial Metaphor. This screen of a library demonstrates the use of a spatial metaphor, in which exploration of a physical space functions both as the central activity for users and an organizational tool for the designer. (*Introducing Media Band,* Canter Technology.)

been the notion of place. Houses, islands, maps, libraries (see Figure 3.1), planets, caves, mazes, studios, galleries, kitchens, and habitats of all types—all these serve not as mere backdrops but as tools for the presentation and arrangement of information.

The most common way for such programs to be arranged is in a hierarchical fashion, in which the first thing the user sees is an overview, or map, of the territory to be investigated. Imagine for a moment a simple interactive program aimed at children and taking as its metaphor the family house. Its opening screen might present a schematic, or cutaway view of the house. A child can then enter particular rooms by using the mouse—by dragging the pointer and clicking. Once inside the room, the child can examine objects, click on them, hear their names pronounced, watch animations, and encounter new characters. This type of arrangement, described here in its simplest form, is an extremely common use of spatial metaphor for the presentation of information. It involves the use of a map or floor plan, or some styled version thereof, as a schematic guide to the information. Users then enter the various rooms to examine the information in detail.

In another variant of this same technique, often used in entertainment programs, the overview is deliberately made inaccessible. On the opening screen, for example, the user might find herself staring down a pathway leading toward a cottage. Only after moving from screen to screen does she discover the perimeters of the environment. This maze-like structure, inviting exploration but not revealing context, has been used with great commercial success in numerous titles, of which perhaps the best known is *Myst*.

Spatial metaphors can be places, large or small, and there may be many objects within that world. These objects—pens, vases, pictures, whatever—may function simply as scenic backdrops, but they may also be vessels of information or tools for the manipulation of the computer environment. (Usually, however, tools are presented to the user in discrete ways, so that their function is apparent.)

It is also possible for an object to function as an organizing metaphor. The book, for example, is one of the most common objects rendered on the multimedia screen (see Figure 3.2). It is frequently used as a design metaphor for the arrangement of information, and with good reason. The design conventions of a book—the contents page, the index, the arrangement of information on a page, the users' ability to write notes in the margins or to mark their place by using paper clips or dog-earing pages, the physical action of turning from page to page—are features with which users are familiar, and ones that are relatively easy to mimic in a multimedia environment. Voyager, one of the early multimedia companies, has in fact built much of its library using the book itself as a design metaphor.

Not all subject matters lend themselves so easily to adaptation through spatial metaphors. Writers and designers need to keep in mind that every metaphor has its limitations. In addition, metaphors can have unintended connotations that infuse the information with inappropriate meaning. In a program for aspiring doctors, for example, it might seem useful to construct a metaphor comparing the human body to a machine, especially when explaining certain physiological phenomena, such as the pump-like action of the heart. Such comparisons, however,

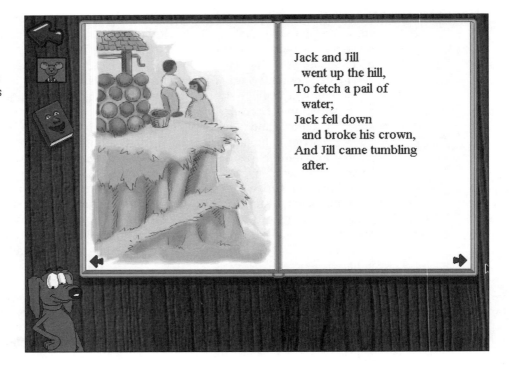

Figure 3.2
An Object as Metaphor. One of the most popular design metaphors is the book, as it offers users a familiar, intuitive interface. (*Jumpstart First Grade,* Knowledge Adventure.)

overlook the spiritual and psychological influences on health, as well as the fact that most patients do not regard themselves as machines anymore than doctors regard themselves as mechanics. The task of the designer, then, is not only to build a design shell that is usable and functional, but also to do so in a way that is appropriate to the subject matter at hand.

Event Metaphors   Another type of design metaphor, called an *event metaphor*, is structured not around movement in space, but movement through time. In such a circumstance, the design shell is structured around an event or series of events —an action with a beginning, middle, and end that moves forward in time. Perhaps the most common example of event-related structures exists in the so-called *twitch games*, the interactive video games that involve the rapid manipulation of a cursor. In these games, the primary metaphor is often a battle, and the primary concern is the destruction of the enemy.

Event-oriented metaphors are also common in simulations, such as flight-training programs or arcade race games, which require the user to maneuver through a series of potential hazards. Although based on an event, these programs also have a strong spatial element.

A more narrative type of event-oriented metaphor underlies some of the so-called interactive cinema. In this genre, the user is invited to participate in an event, such as a senate race, a courtroom drama (see Figure 3.3), a dating game, a criminal investigation, or a reenactment of a famous battle. In such structures, the user typically has a series of tasks to perform and is given tools, or an inventory, to assist in the accomplishment of those tasks. One of the important crite-

Figure 3.3
Use of Characters in
an Event Metaphor.
The action in *In the
1st Degree* is
organized around the
events of a murder
trial, in which the
prosecuting attorney
moves from a pretrial
investigation to the
trial itself. This
program also involves
strong interaction with
individual characters
on a screen-by-screen
basis. *(In the 1st
Degree,* Broderbund.)

ria for the successful design of an event-related design shell is that the event be logically divisible into discrete segments in which the user can participate.

As with most elements of the creative process, design metaphors can often be difficult to categorize. Rarely are they either purely spatial in nature or purely event-oriented, for human events take place in both space and time. Interaction with a computer is a human event, so the design shell inevitably mimics this experience. Moreover, as a design metaphor becomes more richly developed, it often involves not just space and time, but characters as well.

The Use of Characters in Design Metaphors In character-oriented design shells, the interaction with characters is the primary organizing tool. As previously mentioned, characters—particularly human beings—usually exist in time and space, so such designs inevitably become more complex and difficult to categorize. At their most sophisticated, they begin to take on some of the elements of the cinema (though it is important to remember that this is not cinema, and the audience's experience and expectations are extremely different).[5]

Two-dimensional characters have existed on the computer screen for some time, most often as hopping frogs, ninja warriors, and other animated representations of the cursor used in twitch games. One of the first efforts to use characters in a more sophisticated way was in the "Guides Project" developed at Apple Computer. This experimental project featured actors in costume who acted as guide figures, informing the user about particular periods of American history. The concept of guides has continued to evolve. Guide figures are often used now to give a human face to the Help section, providing instructions on how to get around the computer. Another role is roughly analogous to a television host who introduces material and provides narrative segues from one segment of a show to the next (see Figure 3.4).

Another, more challenging approach is to build the design around direct interaction with on-screen characters. Such an approach usually involves some sort of role-playing by the user. *In the 1st Degree*, for example, makes use of an event metaphor, but it is also a character-driven program in which the user plays the part of a prosecuting attorney who interacts with four main characters. The key to the game lies in the successful interaction with the other characters. The same fundamental principle underlies *Brothers*, a multimedia kiosk whose goal is to teach users about HIV, the virus that causes AIDS.

The use of characters in multimedia inevitably starts to raise questions regarding setting and narrative and involves the construction of complex metaphoric worlds. We will address some of these issues in more depth in Chapters 5 and 6.

CLASSIFICATION SCHEMES

The application of a metaphor over an entire environment encompassed by a design shell is a difficult thing to accomplish. It is not always possible or even desirable. Sometimes the metaphor breaks down, and this causes confusion. What had initially seemed such a boon to both user and designer is suddenly a hindrance. Other times the metaphor can get in the way of the message. There

Figure 3.4
Use of a Guide Figure. Marvin Minsky, pictured in the lower left, acts as a guide to his own CD-ROM, speaking to users through QuickTime movies in which he elaborates upon the ideas expressed in the on-screen text. (*Society of Mind,* The Voyager Company.)

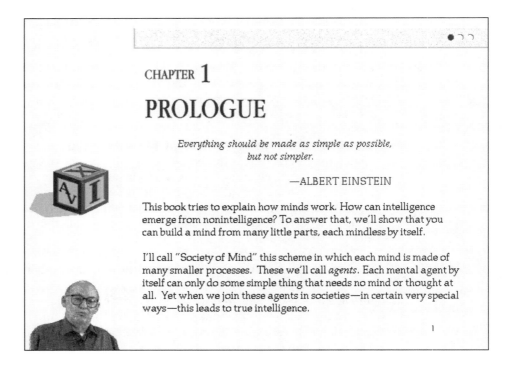

CHAPTER 1

PROLOGUE

Everything should be made as simple as possible,
but not simpler.

—ALBERT EINSTEIN

This book tries to explain how minds work. How can intelligence emerge from nonintelligence? To answer that, we'll show that you can build a mind from many little parts, each mindless by itself.

I'll call "Society of Mind" this scheme in which each mind is made of many smaller processes. These we'll call *agents*. Each mental agent by itself can only do some simple thing that needs no mind or thought at all. Yet when we join these agents in societies—in certain very special ways—this leads to true intelligence.

1

are situations, too, in which the use of metaphor is inappropriate for the audience or for the program's intended function.

In such instances, the designer might decide it is best to get the user into the material as quickly and cleanly as possible, without the trappings of an elaborate metaphor. When this is the case, the primary cognitive tool for the organization of information is *classification*—the sorting of things into categories according to attribute.

There is, of course, nothing new about classification. It is a rhetorical tool discussed by the ancients—by Aristotle in his *Rhetoric* and Cicero in the *Rhetorica Ad Herennium*—and it can also be seen in such mundane places as the daily newspaper, where stories are assigned to different sections of the paper according to topic: World and National News, Local News, Business, Sports, and so on. Classification is a technique that is used, consciously or not, whenever information is organized, broken down into units, and presented to an audience.

The important principle in classification schemes is the selection of the attribute by which the material is organized. The end result most designers strive for are categories that are parallel and without too much overlap of content, presented in such a way that users can quickly get to the information they are seeking.

As an example, consider the case of a writer/designer faced with the task of conceptualizing a design for a multimedia database to be used by veterinarians as an aid in the treatment of animals. Suppose, too, that the writer has already decided to avoid the use of metaphor and to focus instead on a more direct approach.

The writer is then faced with the problem of how to classify the material. In other words, what attribute would be the most useful as the criterion for sorting the material?

1. Should the material be sorted according to whether two-legged or four-legged animals are involved, and then subdivided into additional categories after that?

2. Should the designer eschew physical characteristics, and sort the material by medical procedure?

3. Should the material instead be organized according to symptoms of disease the veterinarian is likely to confront?

The task of the designer is to answer such questions. To do so, the designer must determine the context in which the program is to be used and arrange the classification scheme so that it is maximally useful to its users. For the veterinary database, the designer might decide to sort first by type of animal and then to sort secondarily by symptoms.

In addition to sorting by attributes in ways that are consistent, it is equally important to sort at the same level of specificity. Imagine, for a moment, a user confronted with the following menu choices:

Two-legged Animals

Four-legged Animals

Sharks

The user who had come to the program looking to treat a particular breed of canine—a collie, for example—would know without too much thought to enter the general category domain entitled "Four-legged Animals." However, a different veterinarian seeking to treat a pet dolphin might be somewhat confused, because there is no general category for legless animals. The third choice in the classification scheme represents a higher level of specificity and confuses the user.

Another related consideration is whether to classify information by topic or by type. When classifying by topic, the material is broken up and arranged according to subject matter; when classifying by type, the information is grouped by the medium or format in which it occurs. For example, the home page for Dade County Public Schools Web Site (see Figure 3.5) classifies by topic, providing one category for information on the district, a second category for administrative Web sites, a third for school Web sites, and so on. In contrast, the American Memory Web Site (see Figure 3.6) classifies information according to the type of medium in which the information occurs: prints and photographs, documents, motion pictures, or sound recordings. The decision whether to group by topic or by type is an important one and is based on how users need to access the information.

It may seem that we are belaboring the obvious here, and indeed we are. The truth is, though, that good classification schemes are not so easy to devise as it might seem, and mistakes are commonplace.

We have discussed only a few types of classification. We will discuss others in Chapter 5, including process analysis and comparison/contrast. For now, a gen-

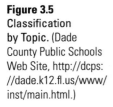

Figure 3.5
Classification by Topic. (Dade County Public Schools Web Site, http://dcps://dade.k12.fl.us/www/inst/main.html.)

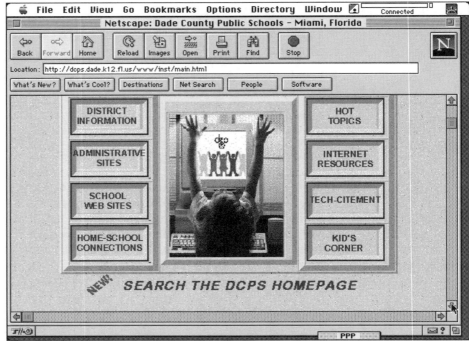

Figure 3.6
Classification
by Type. (American
Memory Web Site,
Library of Congress.
http//rs6.loc.gov/
amhome.html.)

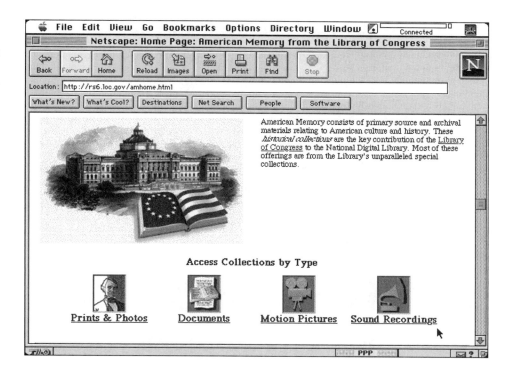

eral rule to keep in mind is that most designers do not necessarily work in a sheerly analytic fashion, memorizing classifications schemes and then applying them to the material. Instead, they first examine the material itself, the task at hand, and the program's intended use; then they search for natural fissures, the junctures at which the material seems to divide itself naturally.

Like design by metaphor, design by classification also has drawbacks. Not all material lends itself to classification, and almost every classification scheme involves some overlap. Establishing equal and parallel categories can not only be difficult but misleading, particularly when it comes to examining subtle relationships among closely related phenomena. Also, classification schemes that are perfectly logical on paper may become unwieldy and difficult to navigate when transferred to the computer.

This is why real-life design so often involves a combination of approaches, using both metaphor and classification schemes (see Figure 3.7). Think, for example, of the Macintosh interface. Although designed in imitation of the desktop, with its documents and folders and trash can, it also uses a simple classification scheme in the form of an option bar running across the top of the screen that presents menu choices according to category. Thus the Macintosh—which has won wide praise for its "user-friendly" environment—makes use of multiple design techniques in a single environment.

A great many writers and designers protest the tyranny of the overriding metaphor, claiming that the computer itself provides the user with environmental

Figure 3.7
An Example of a Dual Design Shell Scheme. In this combined use of a design metaphor and a classification scheme, users can access information one of two ways: by clicking on the categories listed on the note or by clicking on the items in the lunch box. (*MECC 1996 Sampler,* MECC.)

context. When you are working on a computer, the argument goes, you know where you are. What holds the design shell together is not metaphor but consistency of graphic design, typography, and other conventions. Still other designers take this argument one step further, saying that there is no more need for consistency as we move from place to place on the computer than there is during television viewing, in which environments vary wildly as we surf from show to show.

FLOWCHARTS

A common tool used by writers and designers in the development of the design shell is the flowchart. Such charts can be startlingly simple or exquisitely complex, but regardless they are used by most designers as a way of visualizing the interactive structure.

Figure 3.8 is a simple flowchart, based on a classification scheme, that shows the major content areas in a typical interactive brochure of the type that a software company might use as a promotional tool. The boxes represent the material present on a given screen; the lines show the ways in which the different screens are connected.

Flowcharts can be used as a general overview of a program, or they can be quite specific, mapping out a particular branch in great detail. Regardless, the purpose of a flowchart is to provide a graphic model showing the relationships between different blocks of content. A flowchart is in many ways nothing more

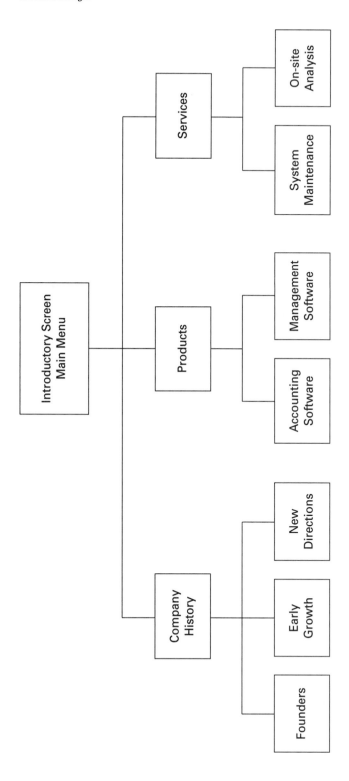

Figure 3.8
Simple Flowchart Depicting the Structure of a
Software Company's Interactive Brochure.

than a visually oriented modification of the traditional content outline that writers have been putting together for centuries. Consider the following outline of the content presented in Figure 3.8:

I. Introduction/Overview
    A. Company History
        1. Founders
        2. Early Growth
        3. New Directors
    B. Products
        1. Accounting Software
        2. Management Software
    C. Services
        1. System Maintenance
        2. On-site Analysis

This type of rudimentary outline should be familiar to anyone who has worked as a writer. Translating it into a flowchart, at least initially, involves little more than reconfiguring the visual relationships between the elements on the page, so that we end up with the relatively simple chart depicted in Figure 3.8. Maintaining that visual simplicity of structure is not an easy task, particularly when the content requires the addition of categories and subcategories and complex linkages between them.

Interface Design

The interface, as mentioned in Chapter 2, is the place where the computer and the user meet. Interface design involves the on-screen arrangement of various content and navigational elements. More importantly, though, it involves the implications of that arrangement for what the user is able to do at any given time.

Imagine, for a moment, an interface that is nothing more than a blank screen with a cursor flashing at the top. Such an interface might suggest entry of data, but it gives little clue about the nature of those data or what is to be done with them once they are entered. Add a tool bar across the top of that interface and fashion it so that its icons resemble those found on the keyboard of an old-fashioned typewriter, and the user gets the idea: This is an interface designed for the entry and editing of prose, otherwise known as a word-processing program. Similarly, take away the typewriter icons from the computer screen, add an artist's palette instead, and shape the cursor like a paintbrush—and now we have a different interface that suggests a whole different functionality and possible set of interactions.

The matching of the interface to program function is vital to the creation of a good multimedia program. It goes hand-in-hand with shell design and the organization of content. In other words, the development of a design metaphor or classification scheme also involves conceptualizing a suitable interface.

Before continuing, perhaps it is a good idea to distinguish the design shell from the interface, and also to distinguish both of these concepts from the interactive pathways. To make this distinction, think of the design shell as a three-dimensional territory, a country through which you are driving. The interactive pathways then are the highways on which you move through the country, and the interface is what you see in front of you at any given time. The interface includes the current view through the windshield—the particular slice of countryside directly in front of you—but it also includes the car's controls: the steering wheel, the rearview mirror, the dashboard, the gas pedal, and the brake pedal. In other words, the interface is the user's window into the content, including the controls for navigating through and manipulating that content. One of the first things a designer has to decide is whether that window is always going to look the same, or whether instead it is going to change as the content changes.

In a word-processing program, for example, the window (the screen) almost always stays the same. The data inside the window may change, but the window itself is constant, as are the tools for the manipulation of that content. In another type of program, users might find themselves confronted with a series of tasks, so the designer might change the look of the interface to fit the tasks and the different types of information presented.

Three primary concerns have to be considered in interface design: content, task, and navigation.

Content is the raw material through which the user is navigating. The interface designer must determine whether the content is essentially the same throughout, or if instead it changes in fundamental ways. Given the nature of the content, the designer needs to consider whether and how to establish standard interface conventions.

Task includes what the users are going to be doing with the content. Will users simply be viewing the content, or will they manipulate it in some way? Given the answers to these questions, the designer must then decide how to provide the tools users will need, including how, when, and where to make the tools available.

Navigation involves the ways users are going to get around the program. Navigational devices are usually present on the screen at all times, but the designer also needs to consider how the users' situation might change according to their location in the program, so that different strategies can be incorporated for getting users back to home base (or the main menu, as it is usually called).

All these concerns are interrelated (see Figure 3.9). Design decisions about one of them affect not just the interface itself, but the design shell as well. In other words, a brilliant design shell based on an extraordinarily clever design metaphor is not a brilliant design shell at all if it has a poor interface. A good interface needs to grow logically out of the material and its presentation. If it does not, the designer may ultimately have to start over by reexamining the material and coming up with another metaphor or classification scheme.

A lot more could be said about interface design, as many regard it as a distinct discipline involving consideration of graphic elements as well as navigational and content concerns. Approached from the perspective of a writer or interactive designer, the interface provides the avenue into the content. Because the content schemes that underlie the design shell impact the graphic possibilities of the interface, it is important for the writer to understand the kind of constraints graphic artists face when trying to implement a particular design. Considerations

Figure 3.9
The Major Concerns in Interface Design: Content, Task, and Navigation. The content appears in the small window. The user's task is to examine and measure the object in the window using the tools provided along the top of the window. The user navigates by closing the window and returning to the laboratory (back screen), which functions both as a design metaphor and staging area for further investigation of other objects. (*Science Sleuths,* VideoDiscovery.)

such as screen size, the number of icons, and the placement of menus and navigational tools are things of which the writer should be aware. Otherwise, the design team may end up with an interface that is impossible to implement.

Aside from practical implementation, there is one general rule that should be kept in mind when considering the design shell, particularly the interface: *Keep the user first.* This cardinal principle underlies all else we have been discussing here, and this principle should be kept in mind at all times. The program exists for the user, not for the sake of the design. Even if a design is ingenious, intricate, and beautiful, it is still worthless if it does not serve the purpose for which it was created.[6]

Design in the Real World

Successful design involves the integration of form and content. The organization of material and its presentation on the interface are key elements in the initial stages of conceptual design for interactive media. Designers working in the field may or may not use the terminology presented here, but all are conscious of these elements—if not in the theoretical realm, then certainly in the practical.

In the workaday world, other constraints and compromises affect the design process. One of these constraints is time. Writers and designers do not work in a vacuum, immune to deadline. There comes a point in every project when a deadline looms, immutable, fixed by the boss and the budget. At such times the team may go with a particular version of the design shell, even though they suspect that it is flawed in some ways or that there might be a better way to organize and present the material. They just haven't thought of it yet, and time has run out.

Another practical concern is technology. Not all operating systems have the same capabilities to present graphics, for instance, and not all are equal in their navigational possibilities or in the way they allow users to interact with content. Interactive programs delivered over networks, for example, may be far more limited in graphic possibility than are programs delivered by CD-ROM. Moreover, other practical issues—load times and programming and memory considerations—sometimes make the best design solutions difficult to implement.

In addition, institutional limitations peculiar to time, place, and circumstance often exist because of tradition, not logic, and those limitations may need to be respected simply because they have been specified by the project's sponsor. Then, of course, there is the pure illogic of human existence, in which human politics, personalities, stubbornness, passion, or ignorance impose design considerations that are less than ideal.

Even in the purest of environments, successful design is no easy task. It involves reconciliation of contradictory goals, compromise, interdisciplinary communication, and the willingness to abandon the rules when nothing else seems to be working. If there is one notion that deserves to remain sacrosanct, though, it is the notion that the user should be kept first. Even that notion, for better or worse, is sometimes thrown out the window.

SUMMARY

In this chapter we have discussed the design shell and two ways of constructing it: through the use of design metaphors and classification schemes.

Design metaphors are a way of organizing the material by comparing it to something else. In this technique, sometimes called design by symmetry, the designer searches for a similarity of structure or process and seeks to juxtapose concepts that are similar at a deep level. These metaphors may be spatial, event-oriented, or involve the use of characters. In creating classification schemes, the designer breaks the subject matter into categories, sorting according to attribute or function.

A useful tool for creating classification schemes, and for developing design metaphors as well, is the flowchart. Flowcharts help the designer visualize the interactive pathways that make up the underlying structure of the program.

The underlying structure manifests itself to the user on the interface, the place where the computer and user meet. Interface design involves the presentation of the various content and navigational elements on-screen. It both shapes the user's immediate experience and determines which interactive options will be available at any given time. As such, interface design must take into consideration content, task, and navigation.

The underlying principle of all design work is to keep the user first. The designer's task is to present the material in ways that meet the audience's needs and expectations while taking into consideration the technical and social constraints that exist in the real world in which the project will be experienced and used.

EXERCISES

1. Imagine that you have been asked to create a multimedia family history. Assume, for the moment, that you will be designing for a CD-like format capable of holding one-half hour of video, about one hour of sound, 100 photographs of various sizes and shapes, and as much text as you want. First, decide what content you would be likely to include in such a project. Having done that, come up with a design metaphor that you might use as an aid in creating a design shell for the material. Finally, draw a flowchart mapping out the major content elements.

2. Working alone or in groups, come up with two alternative design metaphors for the family history described in Exercise 1. Then compare and contrast each metaphor, evaluating them for effectiveness. Consider the following questions: How far can you develop each metaphor before the design symmetry begins to break down? What are the limitations and advantages of each in regard to the inclusion of content? What do the different flowcharts suggest about the experience users will likely have? What might the interface look like for the different approaches?

3. Create a design approach for the multimedia family history based on a classification scheme rather than a design metaphor. Come up with several classification schemes. Using the same criteria outlined in Exercise 2, determine the advantages and disadvantages of each classification scheme.

4. Using a combination of classification and metaphoric approaches, create the best possible design shell for your multimedia family history.

5. Discuss how the design for your family history might change if you were designing for a different environment with different capabilities—for the Internet or for a floppy disk, for instance.

6. Building on the work you've done in the previous exercises, develop an interface for the family history project. First, try to visualize how content, navigational devices, and tools will be present on the interface. With your flowchart in front of you, consider whether or not the interface will remain consistent throughout the program, or whether it will change as the user navigates to different content areas. Draw three different versions of the interface, and then decide which one works best.

7. Choose one of the following topics for a multimedia project:

   a. Your hometown

   b. The history of the automobile

   c. A gardening program

   d. A children's encyclopedia

   e. Rock and roll in the 1990s

   Develop your own design scheme using either metaphor or classification or a combination of both approaches. First decide on the content to include, then choose a design approach, and finally map out the major content areas in a flowchart. (If you don't like any of the listed topics, make up your own.)

8. Develop an interface for the topic you chose in Exercise 7.

ENDNOTES

1. John Gardner, *The Art of Fiction* (New York: Knopf, 1984), 18–21.
2. Marshall McLuhan, *Understanding Media: The Extensions of Man* (New York: McGraw-Hill, 1964).
3. Thomas D. Erickson, "Interface and the Evolution of Pidgins: Creative Design for the Analytically Inclined," in Brenda Laurel, ed., *The Art of Human–Computer Interface Design* (Menlo Park, CA: Addison-Wesley, 1990), 11–16. See also Joy Mountford, "Tools and Techniques for Creative Design," in *The Art of Human–Computer Interface Design*, 17–30.
4. Ivan Sutherland, "A Head-Mounted Three-Dimensional Display," *Proceedings of the Fall Joint Computer Conference*, 1968, 757–764.
5. Brenda Laurel, "Interface Agents: Metaphors with Character," in *The Art of Human–Computer Interface Design*, 355–366. See also Tim Oren, Gitta Salomon, Kristee Kreitman, and Abbe Don, "Guides: Characterizing the Interface," in *The Art of Human–Computer Interface Design*, 355–366, 367–382.
6. Donald A. Norman, *The Design of Everyday Things* (New York: Doubleday, 1988). See also Donald A. Norman, "Why Interfaces Don't Work," in *The Art of Human–Computer Interface Design*, 209–220.

4

Interactivity

The notion of interactivity is a concept that has bubbled up, seemingly out of nowhere, to infiltrate the mass consciousness these past few years. Originally used to describe a certain approach in the field of education, then appropriated by computer designers, the term has recently been used in discussions of everything from television shows to children's toys.

Teachers are encouraged to promote interactivity in the classroom. Advertisers tout interactive products for education. The world is suddenly full of "interactive, hands-on workshops"—all of this as if teachers have not been promoting conversation in the classroom since Socrates, or as if people have not been gathering in groups to talk things over since our species first began to use language.

Interactivity is a concept that currently carries with it a certain buzz, an importance made deeper by our cultural fascination with technology. As always happens in such cases, once a term gets popularized and the buzz gets loud enough, meaning starts to get distorted.

Fortunately, the buzz almost always dies down, or at least it relocates to a different idea or a new fascination. When that happens, the original concept remains behind, somewhat altered perhaps, but still there for us to contemplate and utilize if it seems useful or to ignore if it does not.

In its simplest terms, the concept of interactivity involves an interplay between different elements of some environment. Various elements exist in proximity with one another—the atmosphere and the Earth, for example—and some kind of relationship exists between them, whether it be long term or short, passive or active, close-up or distant. In the human context, such interaction can be between people or between people and things. Getting people to interact with their environment—to explore it, to touch it, and to talk about it—has long been seen as a primary way of teaching. It is a technique rooted in the way people learn about the world from infancy. Certain schools of education, such as the Montessori method, maintain that children learn as much by doing—by interacting—as they do by reading or studying.

Influenced in the mid-1960s by the educational theories of Jean Piaget and Jerome Bruner, computer technologists and theorists began to study the interactions between computer and user. Their primary motive was to design computers that were more useful and could function as an extension of human consciousness. They discovered that different levels of interactivity, or degrees of interaction, can take place according to a user's need and the sophistication of the technology. We touched on the three levels of interactivity in Chapter 2, but those remarks are worth reviewing and amplifying here as we discuss the subject of interactivity in more depth.

The first and simplest level of interactivity allows a user to control access to the content. The television, for example, is a simple interactive device because it allows us to choose the programs we want to watch, to turn them on and to turn them off. When we add a VCR to the television, we increase the level of interactivity. We add the ability to go backward and forward within a program, to watch particular segments repeatedly, and to interrupt the show and return to it whenever we please. The type of interactivity available from televisions, VCRs, telephones, and numerous other everyday technologies gives users the ability to control access to content but not much else.

A second and more complex level of interactivity allows users to choose their own path through the material. Thus computers enhance interactivity because of their vast storage capabilities and ability to access information on demand. Computers make it possible not only to choose which parts of the content are of interest and to go directly to those parts, but also to replay them in a more useful order, even to edit and store new versions of the original material. An audio CD player is a common, everyday example in that it allows users to access recorded material randomly and to reshuffle it, eliminating some songs and preserving others, and even mixing in the work of other artists.

A third and most intense form of interactivity—which today remains primarily theoretical— allows users to change their environments as they go, thus altering not only their own experience but the experiences of subsequent users. At this level of interaction, the computer responds to users' needs on request and perhaps even anticipates them, and the quality of interaction most closely mimics that between living organisms in that both are continually changing in an evolving relationship. Computer games and educational multimedia come the closest to reaching this level of interactivity.

By thinking in terms of degree of interactivity, an interactive designer can help frame some fundamental issues. How much control will users need? How much control must the program retain if it is to maintain its structural integrity and ultimately its usefulness? The answers to these questions depend upon the type of program and the needs and circumstances of the people who will be using it.

More interactivity is not necessarily better, but it does add a certain psychological component because it calls for users to take action. If users' actions are meaningful and produce meaningful responses, interactivity can be a very powerful tool. It offers the opportunity to engage the audience by bringing them into the program and making them responsible for its outcome. In an educational program, it gives users the chance to learn more actively than by merely watching and listening. But giving users a chance to learn by doing also places an added burden on the designer: If there is not a clear relationship between a user's input and the machine's action, or if the machine responds in ways that are difficult and cumbersome, then users become frustrated. Moreover, if an activity is too simple or calls for a series of rote actions, users will become bored rather than engaged. In all programmed action-response models, there is always the risk of dehumanizing users—of treating them like rats in a maze who must first run the gauntlet and then push a button if they are to reap the rewards. Given the powerful psychological component inherent in interactive media, the designer needs to consider carefully the matter of degree, so that the nature of the interactivity suits the program's task, its audience, and the content being presented.

Elements of Interactive Structure

Before moving on to the larger questions of interactivity, we should first discuss the building blocks of interactive structure. When building anything, it is helpful to know the fundamental structural units and how these units are interconnected.

In this regard, it might be helpful to take as our conceptual analogy the cinema, in which the fundamental dramatic unit is the scene. In cinema, a scene occurs in a specific place at a specific time and involves the actions of one or more characters. When that scene is over, another scene follows, and these scenes build upon one another, in a linear fashion, until the film is over.

The computer, of course, is not the cinema, but it does have an analogous structural unit: the screen.

SCREENS

The term *screen* has its origins in the obvious fact that a person sitting at a computer is, in fact, staring into a screen. When a user executes commands that cause movement from one part of a program to another, the visual and functional environment often changes, sometimes quite radically. In such instances, the user's sensation of moving to a new place is analogous to viewing a new screen.

Designers conceptualize users' movement on the computer as movement between screens. The user opens the program at Screen A, explores it for a while, then executes a command that moves the program to Screen B, and so on.

Although screens are often distinguished by their physical appearance, they are more meaningfully differentiated by the type of material made available in them. The background might be the same and the same icons and navigational tools might be present, but the material to be viewed, manipulated, and used has somehow changed.

When a user is on a given screen, he or she may have a number of available options, but these options essentially boil down to two interactive choices: Either stay here and explore the current screen, or go to a new screen. Although nuances are possible within these two realms, the alternatives open at each screen are basic: Stay or go.

Imagine for a moment a solitary person sitting at a computer, staring into the screen. On the screen, someplace out there in that universe called cyberspace, is a bookshelf (or rather the graphic rendering of a bookshelf), and on that bookshelf are a number of books. Aside from the books on the shelf, the only other thing on this screen is a red sign labeled Exit. Suppose that after exploring this screen for a while our industrious user finds that by dragging the mouse arrow over the books, the books can be opened. The user can read what's inside, look at pictures, maybe even type notes in a little notebook, edit those notes, and then store the notebook on the bookshelf. All during this exploration, though, that red Exit sign remains available.

Those then, are the fundamental choices—to browse in the library or to leave it. Again, stay, or go. But when our user leaves, where does he or she go? And how does our user get there? More importantly, how do designers conceive that motion?

MOVEMENT BETWEEN SCREENS

To get a better idea of this process, let's leave for a moment the imaginary library in cyberspace and return to the concrete world of offices and desktops and a filing tool known as a rolodex, a rather simple device used for storing addresses. It consists of nothing more than a collection of index cards, each one punched with two holes and hung in alphabetical order on a rotating drum so that an address-seeker can mount this entire contraption on the desktop and spin through the cards. The advantage of such a system is that it allows people to flit back and forth through the alphabet with relative ease in their search for names, addresses, and phone numbers. Movement through the cards is essentially linear, from one card to the next, from A to B to C, until finally the user reaches the end of the alphabet and comes full circle back to the beginning.

On the computer, contemporary navigational software mimics this system of information retrieval in many ways—except that on a computer, the movement is not from index card to index card, but from screen to screen. The simplest form of movement in such a system would be linear, but on the computer this is not the only type of linking. It is also possible to establish direct links between Screen A and screens that are not adjacent to it. Now, in moving from Screen A, the user is no longer restricted to going to Screen B but can instead go to any screen to which a link had been established—to Screen M, maybe, or to Screen Q3, if such a screen exists.

To visualize this more completely, return again to the computer screen of that lone bookshelf in the cyberspace library. Our imaginary user has returned too and finds that although the books are still there, the Exit sign is gone. In its place are three new signs:

Office

Home

Theater

At the root level, the user is still faced with the same decision: either browse the library or leave—except that now, if the choice is to leave, there are three places to go. The library screen is linked to three other possible locations, all of them bound to the library but not necessarily to each other.

The process of building links between screens, so that the user is given a choice about where to go next, is called *hyperlinking*. This activity is at the heart of building an interactive structure. The designer has two methods for building such a structure: by embellishing the original screen or by building links to new screens.

When embellishing a screen, the designer adds a second layer consisting of sound, animation, graphics, and/or text. This second layer is not experienced, however, unless users deliberately seek it out. To experience embellishments, users must activate them by selecting menu choices, icons, or hot spots. (Though these terms are often used interchangeably—and a great deal of overlap exists among them—there are some distinctions. Menu choices are category labels that help guide users in making decisions regarding content areas. Icons, in general, are graphic symbols whose function is to help users visualize navigational maneuvers or ways to manipulate content. *Hot spots*, which may be iconographic or verbal in nature, are often used to indicate a hidden or disguised spot on the screen that users search out in order to generate a response from the computer.) When the user clicks on one of these added features, the embellishment then plays—it could be a brief animation, for example—but when it is over, the user is still on the primary screen. This is a form of hyperlinking, even though the user is still in the same place in the program.

In contrast, hyperlinking to new screens moves the user along the chain to a new series of navigational choices. Clicking on a hot spot in such an instance, for example, would transport the user to a new screen with different visual content and new interactive options. Moving the user along from screen to screen is the second method used in building interactive structure.

Designers also have techniques for expanding the screen and creating the illusion of seamless movement through a three-dimensional space (see Figure 4.1). These techniques, however, rely on the same fundamental methodology just described; the designer still builds the environment either by embellishing the immediate surroundings or by providing new places to which to go. Regardless of the technical complexity or scope of the individual program, this methodology remains the basic strategy for developing the larger patterns of interactive structure.

Patterns of Interactive Structure

When screens are linked together, they form patterns. These patterns can be displayed visually on flowcharts. Though the numbers and types of possible patterns are virtually infinite, three major patterns function as building blocks for larger interactive structures.

Figure 4.1
Screens vs. Scenes
vs. Realms.

Even though interactive designers often think in terms of the *screen* as the fundamental unit of interactive design, this is not always the case. Virtual Reality Modeling Language, for example, and other programming techniques blur the distinction between *screens,* making the movement more seamless.

Such modeling techniques extend the boundaries of the screen, so that the player uses a pointer or other cursor device to incrementally change the screen. The effect, in which the scenery changes incrementally, mimics walking; the environment changes as the walker travels forward in space and time. On the computer, these extended screens are referred to as realms. A *realm*, then, is an expanded screen or a group of interconnected screens that a user may access incrementally through some form of visual scrolling.

Scene is an older term inherited from theater and film and is used variously by interactive writers and producers. Most commonly the word refers to a piece of uninterrupted video or animation that a user might encounter inside a particular screen or realm. The same word, though, is sometimes used to describe an orchestrated series of interactions, in which video clips are experienced incrementally in response to the user's actions at key decision points. Interactive conversations—between a user and an on-screen character, for example—often consist of a series of video responses to user input. The sum total of these interactions on a particular screen constitutes a scene.

Whether working with screens or realms, the designer faces the same fundamental structural dynamics. In a given realm—as on a given screen—the user can either explore the immediate area or move to the next realm. Stay or go: this is still the fundamental dynamic.

The first and most obvious way of linking material together is in a linear fashion, with one screen following the next. The second is hierarchical, involving a navigation through a series of levels, usually moving from general to specific. The third is a web-type structure, in which items are linked together in a more circular, or looping, fashion.[1] It is important to keep in mind that in the real world of interactive design, these patterns rarely exist in a pure state. More likely, a given program will use a combination of structural patterns, but usually it is still possible to identify the dominant pattern.

LINEAR PATTERNS

There is a tendency among interactive designers, particularly neophytes, to malign linear patterns, to scoff at them as being old-fashioned. In truth, some of the most successful and admired interactive programs to date have been largely linear in their underlying structure, moving the audience from screen to screen (see Figure 4.2) in much the fashion that a reader moves from page to page in a book.

The reason linear structures are enduring, and will continue to be so, is that they mimic and give order to human perception. The simple act of travel, at its core, is a linear process involving movement forward in time and space. When we walk somewhere, we do so one step at a time, examining what lies in our path as we first go here, then turn there, before eventually arriving at our destination. We

Figure 4.2
A Linear Pattern. In this flowchart demonstrating a linear structure, controlled
movement between screens (rectangles) is represented by lines.

can also move along the same path in the reverse direction, as when we retrace
our steps to return to where we began. In the same way—particularly as members of Western culture—we tend to experience life itself as a linear movement
through time, from birth to youth to mid-life to old age and death. This observation is not to deny other, more circular aspects of perception: the notions of
cycles, as demonstrated by the rotation of the seasons; the associative nature of
human memory; or the sense of timelessness inherent to spiritual and meditative
experience. These too are very real parts of human experience and help create
its richness, the sense that we are more than creatures bound by physical dimension. Even so, the basic desire to lay things out in a straight line, to see events and
relationships in a directly causal fashion, remains a very strong drive in our consciousness. And, as it so happens, our major communications media, especially
books and films, have underlying physical structures that are essentially linear in
nature. A book moves forward from page to page; film moves forward from frame
to frame. Once again, computer software—particularly that used for multimedia
navigation—tends to mimic the physical natures of these other media, moving
the user from screen to screen.

In a strictly linear model, as diagrammed in Figure 4.2, the user starts on
Screen A, experiences the material there, then moves to Screen B, and so on. The
options in strictly linear systems are very limited. The advantage is that the
designer can control the information, presenting it in the order that makes the
most sense and imparts it most clearly. The disadvantage is that the user has very
little control and cannot move around at will to examine the material. Users
cannot jump from Screen A to Screen C without viewing Screen B, even if Screen
B is of no interest to them.

There are, though, ways of remedying this and still maintaining an essentially
linear structure (see Figure 4.3). To build more flexibility into a linear structure,
the designer can provide a main menu screen (or contents screen) from which the
user can access all screens in a basically linear stream. The designer might also
add a page toggle, which allows the user to switch back and forth between
screens, moving forward and backward along the line at will. In addition, the presence of hot spots on an individual screen allows for additional enhancement. Such
a modified linear pattern, typical for children's books, is in fact very widely used.

HIERARCHICAL PATTERNS

Linear patterns often suggest movement forward in a sequence, a kind of
inevitable progression of information or ideas. *Hierarchical patterns*, on the

Figure 4.3
Modified Linear
Structure. In this
arrangement, a
basically linear
structure is made
more flexible by the
addition of a main
menu screen and
linkages between it
and all other screens.
Circles represent
hot spots.

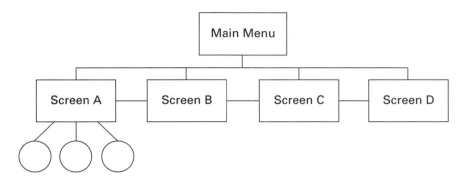

other hand, imply a ranking of different levels of ideas, a movement from general to specific, from set to subset.

Figure 4.4 depicts a simple hierarchical structure. On the first level, the main menu directs users to a second level containing choices. The second level leads to a third, and the third to a fourth. Of course, movement through the program in the opposite direction is also possible.

The advantage of such structures is that they allow users more control, the ability to get to where they want and to find specific information there. The disadvantage is that hierarchical structures tend to be navigation-heavy—that is, users can spend a lot of time searching a pathway for desired information. If the information sought is not found, they must go back to the first level to search alternate pathways.

Hierarchical structures are used in many types of interactive programs, including such everyday utilities as messaging systems, bank machines, on-line services, information data banks, and other applications in which a user's specific goal is to hunt down a piece of information. In such circumstances, the general goal of the designer is to devise a pattern that gets the user to the specific piece of information with the minimal amount of navigation.

Hierarchical systems allow a user a good deal of choice, and they can be flexible to the needs of many different kinds of users. The difficulty is that the bits of information retrieved in this way tend to be disassociated. In this context it can be difficult to maintain the continuity necessary to convey the relationships among ideas. The task of the designer becomes more complex but also more interesting, for each screen must not only have a navigational role but a place in the flow of content, so that there is a progression of ideas as the user moves from one screen to the next.

The important thing to keep in mind is that even in a hierarchical structure, the user's primary experience is linear. Despite the presence of simultaneous options, the user must ultimately decide on only one; the movement is still from one screen to another screen. The job of the writer, then, is to make sure that each step along the way—each link in the chain—makes sense not only in and of itself, but also in the context of what came before and what might come after. When there are many branches and when the branches interconnect, this can be a daunting task.

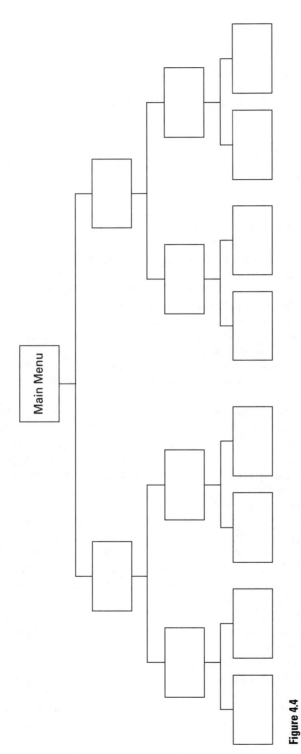

Figure 4.4
A **Hierarchical Pattern**. This flowchart demonstrates movement in a program from set to subset in response to user choices at each level.

WEB PATTERNS

The third primary pattern is the *web pattern*. In a web, the various points are inter-linked in an associative pattern that is neither linear nor hierarchical. One way of visualizing such a pattern is the open web configuration shown in Figure 4.5.

Another web structure, a closed configuration, can be likened to a hollow sphere with various points on its surface, some of which are joined with each other but all of which are joined to the center (see Figure 4.6). This model suggests a universality of access from a central point, as well as a degree of inter-connectivity among the outer points.

Simple web structures are often used to annotate individual screens. Consider for a moment a hypothetical one-screen program designed for children, called "The Music Room," which consists of nothing more than a room strewn with musical instruments. The program's purpose is to teach children the sounds that different instruments make. All the on-screen instruments are "hot" in the sense that initiating contact with any of them triggers a brief musical solo accompanied by an animation showing how the instrument is played. After the solo is over, the user is still on the same screen, inside the Music Room.

Figure 4.5
An Open-ended
Web. This type of
web pattern depicts
the general
relationship of hot
spots (circles) to a
primary screen.

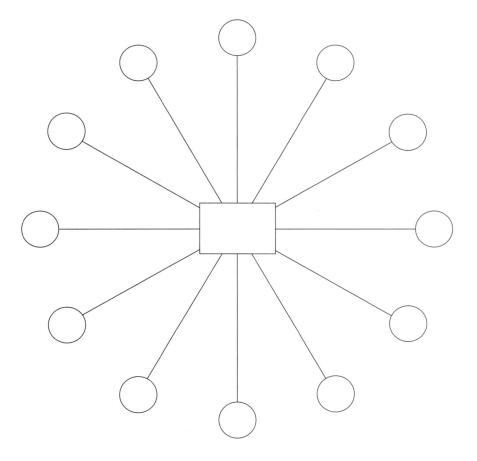

Figure 4.6
A Closed Web. This
kind of web pattern
retains the linkages
present in an open
web but adds
interconnectivity
between peripheral
screens.

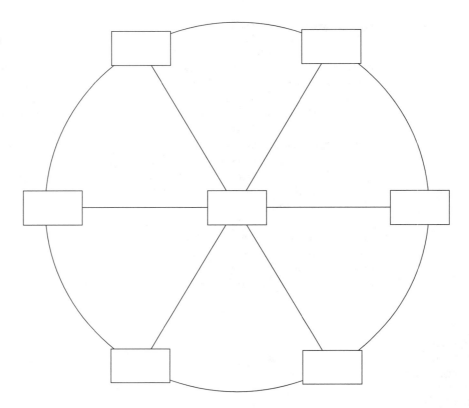

   This type of hyperlinking, in which all hot spots are accessed from a single source, is extremely common. What happens, though, if we decide to build a more complex web? Instead of simply annotating the screen, we could interconnect the web points and use them to transfer the user from screen to screen, so that when the user touches the guitar in the music room, she or he is transported to the guitar room, itself full of various types of guitars from different periods of history. Then, by touching, say, the Spanish guitar, the user now sees a nineteenth-century flamenco dancer accompanied by traditional Spanish guitar. In the guitar room are other choices, ranging from ancient acoustic guitars to modern electric instruments, but there is also a corridor leading away. By travelling down that corridor, the user is suddenly back in the music room, where the journey started. A flowchart of this sort of cross-linking might produce a web structure that is not nearly so symmetrical, but more resembles a spider's web constructed in a midnight frenzy (see Figure 4.7).
   Web structures allow for cross-linking, for exploration across boundaries, for showing the interconnectedness between people and ideas. Used appropriately, webs can get people in and out of information from a central point; they can also be used to create mazes for games and other applications. There are limits, though. Given too much cross-linking, users lose their way. When there are too many interconnections, meaning and control are easily lost, and users may wander in chaos.

Figure 4.7
A Modified Web. The structure depicted here illustrates interactivity in a portion of "The Music Room" program described in the text.

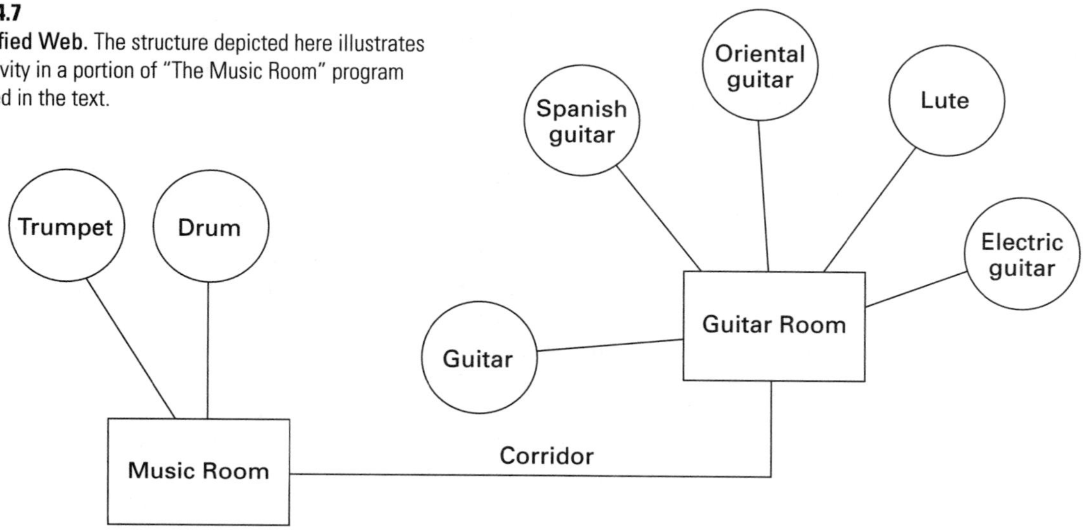

Adding Complexity to Interactive Structure

The three primary branching structures just discussed are intended to serve as descriptive models, not prescriptive ones. They are useful as tools for thinking about design problems, not necessarily as models into which to pour content. In point of fact, designers and writers rarely choose flowchart patterns in advance; rather, they work with the content to see which patterns make sense. They use design metaphors and classification schemes to help chart the content, searching for natural and logical groupings.

The result is that we rarely find the models in their pure forms. Most programs are amalgamations of more than one pattern of interactive structure.[2]

Predominantly linear programs, for example, almost always make use of some hierarchical elements. Similarly, programs that are either primarily web-based or primarily hierarchical often contain elements of the other patterns.

In fact, it is often true that the structure of a completed program, when analyzed and outlined, might look different on paper depending on who does the mapping. In other words, a hierarchical structure can sometimes be drawn to look like a web, and vice versa. If this is true, one might ask, then why bother to use flowcharts at all, let alone make distinctions among different types of structures? The reason is that flowcharts help writers and designers to conceptualize the material, to think about how it can be organized and experienced by the user. Flowcharts can be tools of discovery.

The flowcharts described so far are intended to represent graphically the general shape of the program, but ultimately programs must be charted out in greater detail. How does the writer/designer begin this more detailed process?

As always, the first consideration is the user. The designer must decide whether this is the sort of program users select to find information quickly, or whether it is one in which they will be more willing to spend time, to browse and explore interconnections. Having considered these issues, the designer then must decide whether the interactive structure should be broad and shallow or deep and exploratory.

DEPTH VS. BREADTH

A broad and shallow branch system (see Figure 4.8) is good for accessing information quickly. It keeps everything close to the surface, allowing users to enter it, grab the information, then get out. Such broad branching structures are often used for informational brochures, catalogues, and other programs in which interrelationships are less important than the information itself.

Deep branch systems, on the other hand, although fundamentally hierarchical in nature, can be drawn either on a vertical axis (see Figure 4.9) or on a horizontal axis. Either way, they allow for the extended examination of material and for user choice in regard to that material.

Different areas of a program might use different types of branches, some broad and some shallow, depending on the content. Any branch can be further developed by adding web patterns to the individual screens.

Some techniques for embellishing and developing individual branches are discussed next.

OBLIGATORY CUTS

An *obligatory cut* is a fancy name for a screen or hot spot that presents information that users *must* see. This feature is mandatory viewing; the designer gives users no choice in the matter. Obligatory cuts are used to grab a user's attention and hold it.[3]

The opening screen is the simplest type of obligatory cut. When users boot up a program, the first thing they see is its opening screen, which usually contains some information that the designers regard as essential.

Obligatory cuts can also be used in a situational manner, depending upon a user's location in the program. Suppose, for example, that a user is back in "The

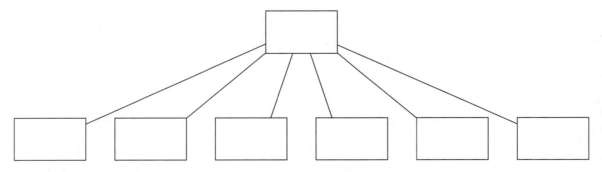

Figure 4.8
A Broad and Shallow Branching Structure. This type of structure is typically used for quick access to information.

Figure 4.9
A Deep Branch.
Deep branch structure
is typically used for
development or
extended examination
of ideas.

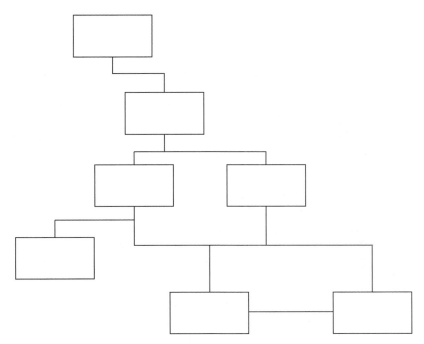

Music Room" program, sampling sounds in a roomful of guitars. As designers, we may have some information—say, about the history of guitars—that we feel is absolutely essential for the audience to know. But there are several ways out of the guitar room, and we can't know which one our imaginary user will take. How can we be sure that the user will see the guitar history lesson? One way is to link the obligatory cut to the room's entrance, so that our user sees it, no matter what, as soon as the room is entered. An alternative way would be to link each exit to the obligatory cut, so that the information would be triggered no matter which way out the user chooses.

The placement of an obligatory cut in either of these locations brings up another problem, however. What if the user reenters and exits the room more than once? Does she have to view the obligatory cut each time? There are several ways of handling this. One is by programming to deactivate the obligatory cut after its first viewing. Another is by building a default path that makes the viewing of the information optional after the first time through. In some cases it might make sense to remove the obligatory cut altogether and integrate the information into the opening screen.

MULTIPLE AND PARALLEL BRANCHES

Programs usually contain multiple branches, each of which may branch one or more times. Some of these branches might dead-end; others might fork out. Some might interconnect with other branches, and still others might "evaporate," in the process returning users to where they started.

Figure 4.10 depicts a hierarchical pattern in which users can choose among multiple branches that move in parallel away from the main screen. Branch A in the figure is an example of a *forking branch*. After a brief interval of linearity, the pathway comes to a fork in the road. At the fork, the user has a choice of directions. In this case, if the user takes the left-hand fork, more options present themselves and the path continues.

Branch B in Figure 4.10, on the other hand, is an example of an *evaporating branch*—one that gives the user a screen or two but then disappears, returning the user to the main menu in the process. Evaporating branches are characterized by the fact that once they have been viewed, they disappear and are no longer available.

Other branches, called *crisscross branches*, are useful when there are points at which the designer deems it suitable for the user to have access to another branch, as if stepping from one boat to another in midstream. This is what happens in Branch C in Figure 4.10. Such crisscrossing can be useful in connecting related ideas, but it can also disrupt the sequential flow of information and may undermine the logical integrity of the individual branches.

Sometimes, of course, branches dead-end. In such cases, the designer must decide what to do with the user. For example, does the user have to execute a command to return to the main menu, or does the program automatically loop the user back to a new beginning? The choices designers make depend on the program's content and structure.

STEPLADDERS AND GLOSSARIES

There are many other ways to structure branches. Figure 4.11 depicts a stepladder branch, in which a continually navigating user is presented with a new choice at each step of the way. Stepladder branches, which continually require users to make decisions, are often used in training and testing situations (and sometimes in games). A correct answer moves the user forward; an incorrect answer ends the chain, stopping forward movement either by returning the user to a previous screen to try again or by returning her to some earlier stage in the program.

Another way of linking material is through a glossary, a data bank of text or images that acts as commentary on the principal screens. Glossaries are in effect reservoirs of reference material. Because glossaries can add so many dimensions to the content, designers often build several ways of accessing glossaries into a single program. The designer might make the glossary accessible through the main navigational menu, which treats the glossary as a destination in and of itself. The designer might also build hyperlinks connecting specific screens to specific parts of the glossary, so that the items in the glossary act as embellishments rather than final destinations. Finally, the designer might build in a search function, which allows the user to call up specified portions of the glossary at will.

A FINAL WORD ON FLOWCHARTS

As previously mentioned several times, there are many ways of drawing flowcharts. There are, in fact, many more structural possibilities than we have dis-

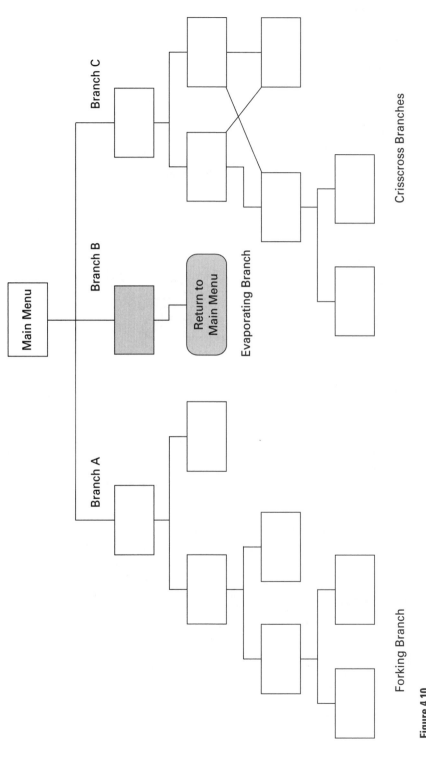

Figure 4.10
Forking, Evaporating, and Crisscross Branches.

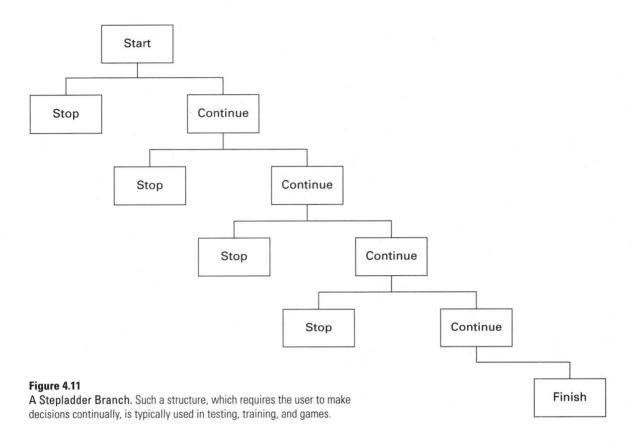

Figure 4.11
A Stepladder Branch. Such a structure, which requires the user to make
decisions continually, is typically used in testing, training, and games.

cussed or illustrated here. Theoretically it would be possible to construct an infinite number of hypothetical variations, naming each and speculating on its possible applications. The truth of the matter, however, is that designers rarely work
this way, as such an approach places form before function. Worse, such extended
analysis of flowcharts in the abstract would soon get tedious and more confus-
ing than it has to be.

This is not to say that inventing structures and experimenting with them does
not have its place. It most certainly does. For practical purposes, though, the primary patterns—web, linear, and hierarchical—are enough to remember. All
other combinations are derivatives of these. In the course of their work, designers invent new variations, most often in direct response to the material with
which they are dealing. Because good interactive programs are designed with the
user in mind, the content needs to be structured to suit the user's purposes. Flowcharts do not exist for their own sake, but as a way of visualizing the flow of content as the user might experience it. To try to design a program by pouring the
content into a preconceived chart would be ill advised—rather like trying to
write a book by diagramming the sentences before they were written and then
plugging in words to fit the diagram.

SUMMARY

Multimedia offers users the chance to interact with the program material, and such interactivity is the primary feature that makes the medium unique. When designed properly, interactivity adds an important psychological dimension by engaging users and giving them the power to act.

Traditionally, designers think of the screen as the fundamental unit of interactive structure, because at any given place in a program, users have two fundamental navigational choices: They can either explore the information available on the current screen, or they can move on to a new screen. Sophisticated programming and virtual reality graphics can blur the boundaries between screens, but the underlying structural principle remains the same.

Writers and designers often create flowcharts to help visualize the relationships among screens. These relationships fall into three major patterns. First, linear patterns contain a sequential linking of screens, one to the next, so that the user is forced to move along a single, central navigational path. Second, hierarchical patterns offer the user two or more mutually exclusive navigational paths at each screen, with each path running more or less parallel to the next. Finally, in web patterns the screens are linked from one to the next in a fashion that allows for cross-referencing and jumping back and forth between several screens.

Linear, hierarchical, and web patterns make up the basic structure for interactive media, though in reality they very rarely exist in a pure state; instead, most programs rely on a combination of the three patterns. Rather than imposing design structures on content, most designers work the other way around. They examine the content in terms of classification schemes and design metaphors, looking for ways in which the content can be naturally and most meaningfully arranged. Of primary consideration, as always, are the needs of the audience.

EXERCISES

1. Choose a multimedia program with which you are familiar (your favorite CD-ROM, computer game, or Web site, for example). Draw a flowchart diagramming the interactive structure. Identify the primary interactive pattern underlying the program. Then reexamine your flowchart, and try to find another way to diagram the same information. Present your analysis to the class.

2. Return to the telephone messaging system you analyzed at the end of Chapter 2. (If you didn't analyze such a system then, analyze one now. Call it over and over, exploring until you understand all its pathways and where they lead; then draw a flowchart.) After you have finished reviewing your initial analysis, identify the primary interactive pattern. Then try to reorganize the system's approach using a fundamentally different pattern. (If their system was primarily hierarchical, use a web pattern, a linear pattern, or some combination.) Is what you come up with better or worse? What are the trade-offs in each approach? Present your analysis to the class.

3. Go back to the family history or other topics you generated in Chapter 3. Reexamine the flowcharts you created. What are the primary interactive patterns? Does one dominate? Now revamp your structure, using a different dominant pattern. Is the end result better or worse? What are the trade-offs?

4. Imagine that your job is to make a multimedia presentation for children that explains how to boil an egg. Create three versions, one using a web structure, another using a linear structure, and the last using a hierarchical structure. What are the flaws and useful attributes of each? Try to construct a combination structure that works better than any of the three primary patterns.

5. Repeat Exercise 4, this time using a topic of your own invention.

E N D N O T E S

1. The underlying notion here regarding the three basic patterns of interactive development was taken in part from a 1994 guest lecture given by Glenn Collyer to my students in the Multimedia Studies Program at San Francisco State University.
2. Many of the ideas here regarding complexities of branching structure I learned while working with Peter Adair and Haney Armstrong, developers of *In the 1st Degree*.
3. The term *obligatory cut* comes from a lecture by Annie Fox, an interactive writer/designer who spoke to my class at SFSU in 1994. The same concept is variously described by other designers as mandatory screen, forced view, and other similar terms.

5

Arranging Content

Because it is the nature of interactive media to offer choices, its audience does not experience a program's content as a continuous flow, but rather discovers it during the process of navigating the various interactive pathways. Much of the designer's job, therefore, involves developing schemes for the integration of content into interactive pathways.

The navigation process in interactive media sometimes relies on *associative logic*: the tendency of the human mind to move from one idea to the next, prompted by seemingly unrelated stimuli in the way that a particular fragrance or color—the smell of coffee in the morning, for example, or a particular hue in the morning sky—may trigger a sequence of apparently unrelated memories: a long-ago trip to Iowa City; a roadside diner; a phone booth containing a woman in a red coat whose face you could not quite glimpse and whom you have never seen again. In such sequences, the mind works imagistically, ordering memories according to sensual and emotional relationships rather than formal logic. In a similar way, a computer user—when activating hot spots, choosing icons, or searching through menu choices—often works associatively, influenced by impulse, juxtaposition, and the internal symbology of the world the designer has created.

Even so, a user's experience is still largely linear. Sitting in front of the screen, the user still interacts with the computer in a sequential fashion, viewing one screen after another and taking one action after another. The user still experiences both the advancement of time and the sensation of movement within the virtual world of the computer, whether that movement be screen-to-screen or realm-to-realm. For this reason, many of the traditional ways of organizing content for film and print also have value in interactive media.

Because this is true, it is useful to look at some of the classical notions of narrative and expository structure. Accordingly, the first half of this chapter examines the traditional narrative elements of story and how they can be integrated into an interactive format. The second half of the chapter discusses expository structures typically used in educational and informational multimedia.

Despite the associative aspect of interactive media, the classical notions of content arrangement we will discuss next are still powerful tools. In fact, together with the primary structural patterns discussed in Chapter 4, these classical notions constitute the basic conceptual tools for writers/designers as they work through the scripting process.

Narrative Structure: Telling a Story

When writers and critics speak of the elements of drama, they are referring to six interrelated elements: plot, character, point of view, setting, style, and theme. These elements are common to all drama, whether it be film, theater, literature, or the daily comics. Though it would be possible to talk about each of these elements at considerable length, we will have to be satisfied with a brief overview here.

PLOT AND CHARACTER

Plot is a term used to describe the events of a story, the incidents that constitute the action, or the sequence in which events occur. Inevitably, writers arrange and present the events of a plot in some kind of order. Often, though not always, the plot has a chronological order (or the illusion of such); the result is that the audience experiences some sense of forward movement, if not in time then in causality.[1]

Aristotle spoke of all stories as having a beginning, middle, and end. Simplistic as that may sound, it remains so sound an observation that for all practical purposes it has proven impossible to deny. Aristotle also observed that most plots begin with the unfolding of a dramatic problem. In other words, stories traditionally begin by presenting to the audience the main character's situation, his or her needs and desires, and the obstacles or complications that stand in that character's way. The next part of the plot, according to Aristotle, involves the rising action, a continuing sequence of events that culminates in a climax, or peak event, in which the character confronts the obstacle most directly. Finally there is the resolution, in which the problem is solved, whether happily or not.

This sequence, admittedly a simplistic account of Aristotle's observations regarding drama, later came to be known as three-act structure. Regardless, once we become aware of these primary narrative elements, we begin to notice them, in seemingly infinite variation, in every format from blockbuster novels to literary short stories to handbooks on how to write for film and television. It is, in fact, such a pervasive structural form for stories that to many writers it is as fundamental and obvious as the observation that trees are made up of branches and leaves. The miracle comes not so much in the simplicity of the basic elements, but in the infinite variation they allow.

Intimately bound to plot is *character*, the people and creatures who inhabit the narrative space and move forward through the story's events. Characters can be conceived of in different ways. There are major characters and minor characters —those at the center of the action, and those off to the side—and then there are protagonists and antagonists. Protagonists are the characters with whom we identify, the heroes and heroines. The antagonists are the adversaries, those who stand in the protagonists' way.

When writers conceive of characters, they typically think in terms of the characters' motivations: their desires and their needs, the things they must do to fulfill themselves. Tied up with these needs, of course, are the obstacles, both interior and exterior, that stand in a character's way. These forces—the desires on the one hand and the obstacles on the other—constitute the push and pull, the dynamics, of a character. These forces are greatly intertwined and closely related to the way in which the character experiences the events in the plot. Aristotle referred to this interrelationship between plot and character as *energeia*. In other words, he saw plot and character together as being shaped simultaneously by a single driving force, so that events seemed to unfold one after the other as the character worked his or her way toward some inevitable fate.

Despite much that has been written to the contrary, plot and character are important elements in multimedia; their manifestation is simply different. This is not to deny a conundrum with which writers and designers must deal continually: If the nature and power of narrative relies on movement forward, on the sense of inevitable motion bound to the vagaries of character, then how do we maintain such power if we give up control to the user? What happens to the *energeia* when the user is allowed to interrupt in mid-sentence or send the character bounding along some other sequence of events?

One way designers have dealt with this is by maintaining a straightforward, linear structure. An effective example is Pedro Meyer's *I Photograph to Remember*, a CD-ROM chronicling the last years of his parents' lives. The program is essentially a slide-show presentation (see Figure 5.1) of photographs arranged in chronological order, with a voice-over narration by Meyer. It is held together by the story line and the compelling nature of the photographs. Some might argue that the presentation is so linear that it would be equally effective on video or film, but this misses the point—that the presentation is capable of producing a powerful impact on the computer screen using an essentially linear structure. The user's proximity to the screen and involvement in advancing the presentation via the mouse add an intimacy to the work that cannot be achieved in other media.

Simple, elegant structures modeled after the linear experience of cinema and books are one vehicle for placing plot and character in an interactive environment. In such structures, the designer's task is to break the story into segments and then decide which material is to be experienced on each screen. In other words, the writer/designer must translate the scenic units of cinematic movement into the screen-by-screen movement that takes place on the computer. In this context, the designer must also decide which material is primary and may require presentation in obligatory cuts, and which is secondary and can be

Figure 5.1
Interactive
CD-ROM Using
Linear Structure.
Photograph from
the slide show
presentation in Pedro
Meyer's *I Photograph
to Remember,* a
CD-ROM with a
linear structure. (The
Voyager Company.)

relegated to hot spots, side trips, glossaries, and other embellishments of primary screens.

Another approach is a more hierarchical structure, in which users are given options regarding the primary path through the material. This technique is useful when the goal is to provide alternative scenarios: to tell the same story from several angles, for example, or to create a single story with multiple endings. In such situations the various paths are sometimes referred to as *story trees.* The designer's task is to arrange background and character material in such a way that the story makes sense no matter which path users travel, even in the event that a user criss-crosses from path to path in midstream. This approach is clearly one of the most difficult ways of constructing narrative and is the method used in *In the 1st Degree.* (See Appendix A.)

It is also possible, of course, to create stories with web-based patterns, structures in which events are interconnected and explored in a more or less circular, free-form fashion. This can make for an interesting collage effect, but the difficulty is to make the exploration meaningful, and this almost always involves some channeling of the user through paths such that events and characters relate to one another in some preconceived fashion. In its purest form this sort of exploration is by its very nature antithetical to narrative, as perhaps best personified by *Puppet Motel,* (see Figure 5.2) a CD-ROM by avant-garde performance artist and musician Laurie Anderson. In *Puppet Motel,* users explore an environment in which randomness and isolation are not simply structural components but part of the overriding theme.

Figure 5.2
Juxtaposition
of the Unexpected.
Musician and
performance artist
Laurie Anderson uses
Futurist techniques
of collage and
juxtaposition.
(*Puppet Motel,* The
Voyager Company.)

No matter the medium, writers need to know their characters, and many of the underlying techniques for developing character are constant. What varies is the way the characters are experienced by the audience. Novelists not only describe characters from the outside but can also delve into their innermost thoughts and feelings, so that the reader experiences the world through the prisms of those characters' points of view. In film, which is a visual medium, directors are unable to make these inward journeys in the same ways and must instead rely on voice-over, facial expression, conversation, and action to portray the inner life. When working with character in an interactive medium, writers lose some of the familiar ways of presenting character but gain some new ones.

As always, in the serious investigation of character, it is the writer's job to know the innermost fears and desires of characters, to understand the kernel at the center of their being, what it is that drives them forward into action or freezes them in inaction. In other words, the writer must determine each character's motivation. Given that multimedia is largely a visual medium in which users often experience text in an ancillary fashion only, the writer must find ways to establish motivation other than the inward exploration novelists use. Similarly, the visual grammar of film, in which character is developed through immersion in activity, is also changed. Action is still possible, but it is rarely the extended, uninterrupted action of cinema. Instead, action occurs in short discrete bits.

So the traditional ways in which character manifests itself to the audience are altered in multimedia. Scenes tend to be brief. Text often takes the form of headlines. Longer documents may be placed off to the side for access at the user's pleasure (if at all). Audio is often short and sweet. None of these observations are rules, but rather characteristics of the how plot and character are typically experienced in an interactive environment.

In multimedia, then, character, much like plot, ends up being created by collage.

The major task facing the multimedia writer is how to convey the psychology of the characters to the user. Toward this end, multimedia writers use certain techniques unique to an interactive environment, all of which spring from the audience's ability to interact with and manipulate the characters within the program.

One way in which character can be developed is through direct conversation between the user and the character. Such conversation is almost never free-form; rather, the user is typically given a choice of questions to ask. The writer needs to create a context for such conversation, and questions need to be framed so that they are more than simply informational. Moreover, the writer must create alternative responses for characters, so that their responses can change with changing circumstances, just as in real conversations. This type of conversation takes advantage of one of the features of the medium: the ability of on-screen characters to directly address users, to seem to look them in the eye and say what is on their minds. Moments when characters address the audience directly are rare in books and film and often come off as intrusive. In interactive media, they are natural.

Another way in which character is developed is through mission: by sending users off on an adventure of discovery or exploration. As part of such adventures, users may act as a kind of god, enlisting the other characters as surrogates on a field of action and then watching their behavior. Alternatively, users may act as participants in the program and engage in a direct, adversarial relationship with the characters on screen. Positioning the user in regard to the other characters is closely related to another familiar narrative element, point of view.

POINT OF VIEW

Traditionally, the notion of *point of view* has had much to do with the angle from which the story is told, or whose perceptions lie at the heart of the story. Novelists make use of many different narrative stances, including the omniscient narrator, in which the narrator knows the thoughts and actions of all the characters; a first-person narrator, who is limited to his or her own observations; and the so-called objective third person, in which the narrator observes characters from the outside without making subjective comments.

In film, these basic narrative approaches remain, but point of view also takes on a more technical aspect having to do with the angle at which the camera is placed and what that angle implies about the perspective from which the story is being told. In multimedia, point of view retains these characteristics, but designers must also consider point of view in answering a fundamental question in the construction of the design shell.

Where does the user stand in relation to the material? There are essentially three possible answers to this question. In the first possibility, the user can be the point-of-view character. In such cases, the user often has a task to perform and may adopt a persona as part of that task. In other words, the user may have a role to play in the unfolding drama—as, say, an attorney, or a cop, or a dragon slayer. In the second possibility, the user may play a more omniscient role, manipulating the characters and the action as if from on high. In the third instance, the user may instead be a watcher, a member of the audience experiencing the program in much the way that TV programs have long been experienced.

Determining the user's relationship to the material is one of the cornerstone decisions in developing any interactive program. In some ways it is not very different from the writer's age-old concern over audience. There is, though, an important difference, particularly when the user acts as a character in the program. In such instances the user has a role to play, and the writer must develop the role of the user's character. In other words, the writer has to impart motivations to the user as well. What it is that drives users might be simple, such as the accumulation of points in an arcade game, or quite complex, as in a business conflict-resolution training program in which the goal is to get diverse employees from diverse backgrounds, each with their own interests and agenda, to work together toward a common goal.

Finally, interactive media lend themselves readily to examination of material from alternating points of view. This is not to say that the use of multiple points of view to tell a single story is somehow unique to multimedia. William Faulkner did it in *As I Lay Dying* (1930), and the technique has its roots in the early days of the novel. The same general observation is true of film and painting, in which artists have experimented with deconstructing a subject matter and then examining it from different angles. Pablo Picasso did this frequently, particularly during his cubist phase, and filmmaker Akira Kurosawa used a similar technique in his famous film, *Rashomon* (1951), which takes one story and retells it from several points of view.

Interactive multimedia lends itself to such Rashomonesque use of point of view by its very nature. As users, we are constantly drawn down alternative paths and away from the center. Often the difficulty is not in finding new paths to explore but in remembering where that center is.

SETTING, STYLE, AND THEME

Setting involves the places in which stories happen. Settings can be very elaborate or extremely simple. They can drive a story—characters can be lured, seduced, even killed by the settings in which they find themselves—and rarely are they mere backdrop. No matter how pared down or elaborate, they make up an important psychological component of any program, regardless of medium. A setting, in fact, can be so powerful as to emerge as a character in and of itself.

This is especially true of computer driven media, given designers' tendency to regard the computer as a spatial medium in which a user can roam and explore. Setting, in fact, is the dominant element in *Myst*, an exploratory multimedia drama set on an imaginary island. In *Myst*, the user explores the island and learns about its former inhabitants and their fate. The exploration of the setting, of the maze-like island and the riddles it presents, serves as the dominant narrative element.

Another important dramatic element is *style*. In fiction, the term refers to the decisions the writer makes at the sentence level—in terms of syntax, diction, and vocabulary—that give the prose its distinctiveness, its vividness, its sense of period, its tone. In film, style generally refers to visual elements having their roots in the art of cinematography. When people speak of a multimedia program's general style, they are often referring to the interface and its blending of textual and visual elements.

There are two primary interface styles. One is an intuitive or direct style, sometimes called a transparent style, in which the designer's goal is to make all navigational choices easily discernible by the user. The contrasting mode is a more explorational style, in which the interface is presented as a puzzle for the user to solve. In this mode, the navigational icons and content buttons are not readily apparent, but rather something for the user to search for and discover. The philosophy behind the explorational style is that the interface should be playful. As in other aspects of design, the designer's choice between transparent or explorational styles has much to do with the program's audience.

The final narrative element is *theme*, which, put simply, is the meaning of the work. A program's thematic statement is a distillation of its meaning, a recantation of its message. When we talk about the theme of a work of art, we are in a sense paraphrasing the work itself, the way in which all the elements come together to produce an effect, a realization, or a moment of catharsis. Artists and writers are often reluctant to talk about theme, not because it isn't there, but because they fear such discussion will limit the theme or over-simplify it, thereby robbing the work of art of its mystery. Even so, audiences feel a great need to partake in such conversation; critics help to satisfy this need by making it their job to scrutinize and analyze the works of others. In academic circles, a rich and complex language for talking about theme has evolved for both film and literature. Some of this discussion makes its way into a variety of periodicals in reviews of contemporary films and books.

Such a vocabulary has not yet emerged in regard to interactive works, probably for a simple reason. To date, most interactive programs—some people would say *all*—are not worthy of serious critical discussion. Many of the works produced so far have been reference works, simple compendia of video and textual files on a given subject. Little of that has been produced with attention to theme or with the intent to reveal something about the human condition. Interactive narratives, for the most part, have been simplistic elemental stories of one person punching or shooting another—a repetitive drama acted out in myriad forms on video games. It could be argued that there is little reason to discuss such things seriously.

Perhaps so, but there is something visceral going on even in the so-called twitch games, perhaps something worth understanding. It may not be long before some artist combines that visceral energy with other narrative elements to create a vitally charged experience demanding critical attention. (Perhaps some work of this nature already exists but has been ignored because computer gaming is not taken seriously in critical circles. This, however, is beginning to change.) In the meantime, a critical vocabulary is just starting to develop.

It is important to note that the various narrative techniques discussed here should not be considered the province of games alone. They have usefulness in the educational arena as well. They are every bit as important in education and business as the expository techniques we discuss next. As we will note in Chapter 8, educational multimedia often relies equally on expository and narrative techniques, blending both in an environment remarkably similar to that of a computer game.

Expository Structure: Conveying Information

Though the use of narrative elements is important in many multimedia presentations, there are other ways to approach structure, particularly in nonfiction works and for educational, training, and sales programs. Primary among these is a contemporary adaptation of an age-old structure long used for rhetorical presentation, argument, and persuasion.

We have mentioned Cicero's *Rhetorica Ad Herennium* before. It is in this handbook—allegedly written by the Roman orator not long before he was executed by the emperor Nero—that the elements of the classical presentation were first laid out. According to Cicero, these elements were the introduction, the division, the narration, the evidence, the refutation, and the conclusion. In the following sections we will see how these various elements might apply to interactive media by discussing the introduction and the division as introductory elements, the narration as background material, the evidence and refutation as the main body of the presentation, and then finally the conclusion.[2]

INTRODUCTORY ELEMENTS

In a classically structured presentation, the *introduction* is that which comes first. It introduces the material, and it also acts as a bridge, a transition between the outer world filled with distractions and the subject matter at hand. The introduction might perform its task in an anecdotal fashion, slyly, casually, or it might do so loudly and forthrightly, immediately stating its goals and intentions. Regardless, its major goals are to capture the audience's attention (and hold it) and to let the members of the audience know what it is they are about to experience. Another purpose, sometimes only hinted at, other times stated directly, is to teach or persuade the audience. No matter whether the persuasion is on a grand scale or far more modest, most good introductions offer a slant, some insight into the material about to be discussed. To put it in old-fashioned terms, they offer— whether explicitly or implicitly—a *thesis* for the audience to consider.

In interactive media, the introduction takes place on the opening screen or in a sequence of opening screens. In the opening it is incumbent upon the writer/ designer to capture the user's attention and to do so in a way that is indicative of both the subject matter and the thesis.

An example of this involves the opening screen to Voyager's educational program entitled *Macbeth* (Figure 5.3). On the face of it, this screen—a stone gray surface featuring the title of this famous play, with a dagger etched upon the background—is extremely simple. But as we watch, the etching of the dagger turns red, as if filling with blood, and then we hear a voice-over: "Present fears are less than horrible imaginings. My thought, whose murder yet is but fantastical, shakes so my single state of man that function is smothered in surmise, and nothing is but what is not."

This simple combination of textual, visual, and audio elements on the opening screen is more than a teaser. It introduces the subject matter in a way that sets the tone for what follows. More importantly, the designers' choice of imagery and words emphasizes what they believe to be the main thrust of the play. After view-

Figure 5.3
Opening Screen.
(*Macbeth,* The
Voyager Company.)

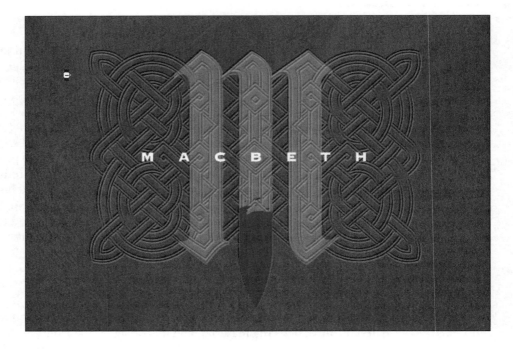

ing the opening screen, the user has the option of entering an extended, formal introduction consisting of a linear slide show, with voice-over, that introduces, discusses, and elaborates on the main thematic elements of the play.

One of the decisions facing a designer is whether or not to have a controlled introduction—whether or not to structure the program so that it forces users through some introductory material, like it or not. If the answer is yes, then a second consideration arises: How long—and how interactive—should that introduction be?

To make this decision, the designer must consider not just the material but the audience. How often is a user likely to use this program, and under what circumstances? If the program is something the user will view once or only once in a great while, then the designer may choose a controlled introduction. If, on the other hand, the program is something people will return to more often, then the designer needs to give users a way to circumvent the introduction. The designers of *Macbeth* opted to solve this problem by providing an opening screen that encapsulated the major thematic concerns of the introduction and then adding a more formal introduction users could enter or exit at will.

The *division* is the next element in a presentation, and it often appears as part of the introduction. In a classroom situation, the division might be that part of a lecture in which the teacher stands in front of the class and says, "Today I want to talk to you about the three principal characters in this play: Macbeth, Lady Macbeth, and Macduff. I will discuss each of them in turn, tracing their moral development through key scenes of the play." In other words, the division is that part of the presentation in which the speaker reveals to the audience, the underlying classification scheme, including what will be discussed and in what order. As a simple graphic aid, the teacher might count off the main points on his or her fingers, or even write headings on a chalkboard. In a similar way, textbook authors use headings and subheadings to guide a reader through the text.

In multimedia, the division is often found on the main menu in the guise of the navigational categories presented to the user on the interface. In the CD-ROM version of *Macbeth*, this division appears immediately after the opening screen, in the form of a table of contents from which the user can decide where to go (see Figure 5.4). Designers frequently use this method and supplement it with a pull-down menu hidden behind a toggle bar at the top of the screen. In other instances, however, the division may be less obvious or may not be there at all, particularly in programs in which the designer wants users to hunt and explore.

BACKGROUND MATERIAL

Another element of presentation is the *narration*, or background material, consisting of information the author considers imperative if the audience is to understand the subject at hand.

The major question the designer must answer in regard to background material is where to put it. Should it be included as an obligatory cut after the introduction? Should it be placed off to the side, in a category of its own? Or should it be accessed through hot spots in the appropriate places as needed throughout the body of the presentation?

Figure 5.4
Division of Contents. This second screen of *Macbeth* functions as both a table of contents and a primary navigational screen. (*Macbeth*, The Voyager Company.)

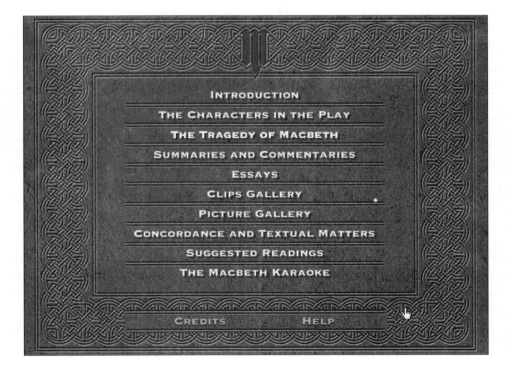

The answer to this question rests partly on the nature of the background material and partly on how essential it is to the user's understanding of what directly follows. The more essential that background material is, the more likely it is to be hardwired into the main presentation, or at least situated where it is difficult to ignore. In the case of *Macbeth*, the designers addressed this problem in two ways: They included background material in the formal introduction, and, they built in a hyperlinked glossary and commentary system that was available screen-by-screen so that users could get background whenever they needed it.

THE MAIN BODY

The main body contains the bulk of the presentation. In rhetorical exposition, it presents the evidence, the examples, the appeals to authority and history relevant to the subject at hand. The main body can also contain a refutation, a special section in which opposing arguments are discussed and then dismantled point by point. The main body provides information that fleshes out the central idea; it is the stuff that substantiates the thesis. In educational programs like Voyager's *Macbeth*, the body of the program is the text of the play itself, complete with hypertext annotations, hot-linked video clips, glossaries, scene commentaries, and analysis.

Most expository presentations have a rhetorical purpose, whether it is to get the audience to look at an old subject in a new way, to convince them of the validity of certain argument, or to buy a product. To organize the material in the body of the presentation, writers and designers rely on familiar rhetorical techniques. Primary among these is classification—the breaking of material into categories, each of which pertains to and supports a central idea or notion. In addition to the general classification schemes we discussed in Chapter 3, of particular relevance here is the classification scheme called process analysis.

Process analysis does just what the name implies: It analyzes a process. It does so by breaking a process into stages, defining them, ordering them, and then presenting them one at a time so the audience can examine them in isolation and understand the relationships between them. Process analysis is useful in training programs and for general educational purposes.

The process of growing a vegetable garden, for example, might be divided into five steps: planning, planting, growing, harvesting, and caring for the garden during the off-season. These steps could be broken down further. The planning stage, for example, has its own discrete steps involving research into the climate, the suitability of various crop species to local conditions, the allotment of available garden space, and the purchase of seeds and equipment.

The advantage of interactive media in presenting process analysis is that it allows users to access those parts of the process in which they are most interested. Accompanied by cutaway graphics and videos, process analysis can be an invaluable learning tool, one that is already used by military and corporate trainers. From an educational standpoint, it is not difficult to visualize how this technique might be useful in teaching everything from language skills to the physiology of plants. To dispel any doubt about the efficacy of teaching through process analysis, we need only think of the now-ubiquitous ATM and how it has trained the public to act as their own tellers, guiding them through the process of accessing the bank's computer and making deposits, withdrawals, and transfers.

Another useful form of classification is *comparison/contrast*, in which two or more objects are compared for their similarities and contrasted for their differences. There are two fundamental organizational patterns for using this technique.

The first possibility, depicted in Figure 5.5, displays comparison by subject. Here there is a category for each object to be compared. In this hypothetical program, the designer's task is to enable comparison of the characteristics of different breeds of dog. In this structure, the user navigates from the main menu to the Shepherds screen and then examines the animals' various characteristics from that location.

The second possibility, diagrammed in Figure 5.6, displays comparison by attribute. In this case, the dogs are not classified by breed; instead, attributes— in this case, size, temperament, and breeding—form the bases of comparison. In this structure, the user navigates from the main menu to the Size screen and then examines the available canines from that location.

One of the special capabilities of interactive media is displaying information in windows. This feature enables users to call up new information in the window while keeping the old information on screen, allowing the side-by-side comparison of both. The hyperlinking function is also a valuable tool in drawing comparisons.

In addition to classification, process analysis, and comparison/contrast, a number of other rhetorical techniques are effective. Among these are the use of example, which involves developing hot links to provide explanations and illustrations to support specific ideas. Another is definition, which is commonly accomplished through glossaries. Evaluation and testing functions are often built into educational programs, sometimes with multiple choice and other times via more open-ended questions. Cause and effect relationships can be developed through the use of nested windows. All these rhetorical techniques and others— including the appeal to authority, elimination, negation, refutation, and explication—can flesh out branches and develop hot spots and can also function as the bases for classification schemes.

Figure 5.5
Comparison by Subject.

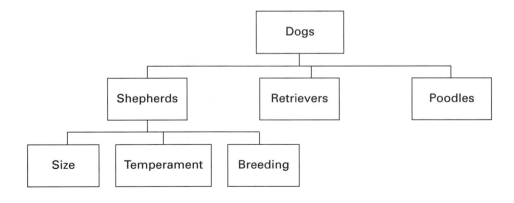

Figure 5.6
Comparison by
Attribute.

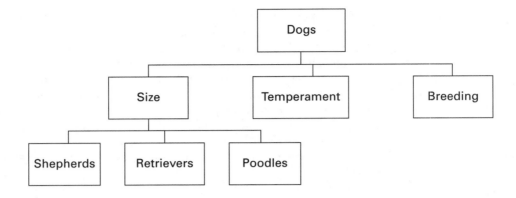

CONCLUDING ELEMENTS

The final part of a presentation is the conclusion. In classical presentation style, three things typically happen in a conclusion: summarization of the main points of the presentation; reiteration of the thesis, usually in some emphatic way; and a call-to-action, an encouragement to users to go out and do something with the information they just learned.

These same features generally apply to conclusions in interactive presentations, but accomplishing them can be a sticky matter, because the writer/designer can never be sure exactly when a particular user is going to leave the program, or which particular route that user will have taken to reach the conclusion.

One solution is to leave exit options open-ended so users can opt for a quick exit. This allows users to return later to view the conclusion when they are ready. In training pieces, designers sometimes include a concluding summation that trainees can come back to and review periodically.

Another approach, the opposite extreme of the previous solution, is to force viewers through the conclusion, no matter what. This approach is often used in sales pitches, for example, when the intention is to offer potential customers one last look at the product or a last chance to fill out an order form or call a toll free number.

A third approach is to build a series of alternative conclusions for each of the paths that users may take. This involves, of course, some careful construction of all the possible decision paths. If there are, say, three main content areas, and a user navigates one of them but not the other two, then he or she would see the conclusion for that area only. Whenever users are able to navigate back and forth between branches, however, the structuring of conclusions in this manner becomes far more difficult.

Conclusions can be elaborate, involving several screens, or they can be simple and concise, involving no more than a tag line or slogan that somehow brings it all together.

Keep in mind that in the building of conclusions, multimedia writers and designers have the advantage of having the last word in a way that may not exist in other media. A reader can escape a book before it ends simply by closing it; a viewer can escape a movie before the last scene by walking out the door. On most computer media, however, the user has to shut down the program before leaving, and thus the designer can always show the user one last screen or series of screens before the user can escape. The use of these closing screens is a powerful tool that can be used to communicate the key parts of any message.

SUMMARY

Traditional narrative and expository techniques have important places in interactive design, particularly because most interactive pathways have a strong linear structure. Given this fact, writers and designers often use classical notions of content arrangement to help compose the linear, hierarchical, and web patterns that make up the underlying structures of interactive programs.

The elements of classical narrative can have as much relevance in multimedia as they do in print, film, and theater. These elements include plot, character, point of view, setting, theme, and style. Though their manifestation in interactive media may be different, the underlying aspects of these fundamental elements remain much the same. Plot elements, for example, can be adapted to linear and hierarchical patterns, or they can be revealed in a more pastiche-like fashion through the use of a hyperlinked web. Whatever interactive structure is used, though, plot elements must still be conceived by the writer/designer. Such conceptions are almost always made, whether consciously or unconsciously, accordingly to the age-old precepts of storytelling.

Similarly, classical rhetorical techniques have their place in interactive design. The traditional expository work has these basic parts: introduction, background, main body, and conclusion. Even in the classical expository model, however, not all of these parts are used in every essay or speech because the arrangement of these elements must be flexible enough to match the audience and subject matter. For this reason, classical observations regarding expository presentation are extremely useful and easily applied to interactive media.

According to the classical notions about introductions, for example, a speaker's opening statement has three functions: to catch the listener's attention, to introduce the subject matter, and to introduce the thesis, or main point, of all that is to follow. These same observations can be adapted and applied to the opening sequence of interactive presentations, particularly those designed for education and commerce. Similarly, interactive writers and designers need to provide background for their audience, integrating into the presentation specialized information, terms, and ideas that might be needed to understand the thrust of the program. Such backgrounding may occur in subscreens, glossaries, or special menu sections. In the main body of the interactive presentation, rhetorical techniques for arranging information—such as process analysis, comparison/contrast, and refutation—hold as much validity as they do in other media, providing a means for developing and supporting the main idea. The basic functions of the conclusion, too, remain the same: to summarize the main points, to reinforce the thesis, and to move the audience to action.

The strength of such classical notions lies in the fact that they are descriptions, not prescriptions. Their utility is not in saying what must be done, but rather in developing a conceptual method for understanding the basic elements of narrative and expository structure. The usefulness of these elements rests in their flexibility and in their seemingly infinite combinations. For these reasons, they are excellent conceptual tools for the interactive writer.

EXERCISES

1. Go to the library or a bookstore, and find a children's book that particularly interests you. Analyze the book in terms of its narrative elements—plot, character, point of view, and so forth—and then come up with a strategy for adapting this book to interactive media. (Hint: One approach might be to do a scene-by-scene analysis of the essential scenes, and then converting each to a screen. The screens might be linked together in a primarily linear fashion and embellished with hot links. Try this pattern, and then experiment with adapting the story using hierarchical and web patterns.)

2. Consider a family story or a fairy tale that you heard often when you were growing up. Write that story down, as completely as possible, and then analyze the story in terms of its various narrative elements. Do a scene-by-scene breakdown, and construct a flowchart that adapts the story to interactive media using a primarily linear structure.

3. Reread the first part of this chapter, and then experiment with different ways you might tell the stories you adapted in Exercise 1 or 2. (Hint: Try building a structure that examines a single story from alternative points of view, or try a place-centered approach in which the user discovers the story and characters by exploring a setting. For either of these approaches, experiment with several combinations of web, hierarchical, and linear structures.)

4. In each of the previous exercises, what is the stance of the user relative to the program? (In other words, is the user a character, a manipulator, or a passive watcher?) What is the theme of each of the different stories you have developed? In what ways might you strengthen the theme without making it silly or reductive?

5. Go to a local health clinic, library, public agency, or political action group to get a brochure on a subject of public interest. Adapt the brochure to multimedia, using the classification schemes implicit in the original brochure. Draw a flowchart that shows how the user might experience the program.

6. Using the same or a different brochure, adapt it again, this time using the classical presentation style outlined in this chapter.

7. Conduct research on a subject of interest to you (local environmental issues, for example). Outline the major components of the subject matter, and list the points you consider important enough to include in your presentation. On paper, organize your material into classic presentation style. Devise an adaptation strategy, using any combination of linear, hierarchical, or web patterns. Draw a flowchart that shows how a user might experience the program.

ENDNOTES

1. Numerous works discuss the elements of narrative. See Aristotle, *Ars Poetica*; Syd Field, *Screenplay: The Foundations of Screenwriting* (New York: Dell, 1979); Jon Franklin, *Writing for Story* (New York; Mentor, 1987); John Gardner, *The Art of Fiction* (New York: Knopf, 1984); Charles E. May, *Short Story Theories* (Athens: Ohio University Press, 1976); and Philip Stevick, ed., *The Theory of the Novel* (New York: Free Press, 1967).
2. For additional information on expository techniques, see Thomas W. Benson and Michael H. Prosser, eds., *Readings in Classical Rhetoric* (Bloomington: Indiana University Press, 1969) and Lane Cooper, ed., *The Rhetoric of Aristotle*, (Englewood Cliffs, NJ: Prentice-Hall, 1932).

6

Scripting and Designing Computer Games

People enjoy games. Playing them is an elemental behavior, childlike and charming on one level but deadly serious on another.

On the surface, games may seem little more than diversionary activities, sheer fun, excuses for adults to reenter a childhood realm and escape the complexities of everyday life. If we plunge a little deeper, though, we see that more is involved than simple escapism. Games, for adults as well as for children, often provide ways of learning about the real world and practicing skills important to human development. They can also provide a context for social interaction.

More important, perhaps, is that games tap into deep primal energies inherent to human experience but often inexpressible in other ways. To be able to compete without sanction, to defeat close friends, to win and lose gloriously, to experience love, betrayal, and death in ritual format—all are possible within the dramatic structure of games. Such experiences are vital to the human psyche and far from trivial.

The pervasiveness of games—in cultures ancient and new, among people of all ages—suggests they are of more than casual importance. On a social level, games are often associated with ritual occasions, just as today's sporting events often occur on key religious and national holidays. On an individual level, there are games for almost every stage of life, from childhood to old age.

Computer games are the most recent heirs to this long tradition of game playing. Like all sorts of games throughout history, they have been reviled as scurrilous and frivolous, as a source of personal and social degeneration. Various studies have blamed computer games for everything from the rising incidence of carpal tunnel syndrome, to declining school performance, to overweight and socially inept teenagers. Such observations no doubt have some elements of truth; even so, computer games still offer good visceral fun. They provide a form of entertainment that can be less passive than watching television, and they also teach computer skills that are becoming increasingly valued in society (even if there is still a part of each of us that wishes computers did not exist).

The bad thing about games is often neither a game itself nor the medium in which it is played. The danger instead comes when people become so involved in games that their lives atrophy and grow stale. Exactly when such a danger becomes imminent, however, is difficult to judge. Adolescent devotion to outdoor games such as baseball and football is by and large considered a healthy thing, and the successful carrying of this devotion into adulthood has spawned a very lucrative industry with highly paid professionals who are greatly admired. A similar industry has sprung up around computer games. This industry, in fact, has for many years been the financial lifeblood of multimedia, serving as a testing ground for technical possibilities, as well as for ideas about interface design and the blending of interactivity and narrative.

Devoted passion for any activity has its dangers, but it is also a characteristic of many realms of human endeavor. What seems pointless and trivial to one person is for others a visceral and rewarding pursuit that informs and enriches their lives.

In their efforts to make the transition to interactive media, many entertainment writers find themselves at odds with the culture of computer game development. This is partly because game development has been the province of programmers, and hard-core game enthusiasts have a natural suspicion toward the writer's emphasis on narrative. Moreover, there is a tendency among traditionally trained writers to look down on computer games as frivolous activities. Both attitudes are mistaken, as game developers and more traditional writers alike have much to offer one another.

Writers interested in multimedia would do well to take a hard look at games, particularly when it comes to matters of interface development and advanced techniques for engaging users in an interactive environment. These techniques are applicable not just for computer games, but for other forms of interactive media as well, including educational and business programs.

This chapter begins by examining the basic skills and material elements common to all games and then discusses the three major types of computer games. Finally, we will take a look at the scripting process for games, including some excerpts and samples.

The Nature of Games

The computer game industry tends to classify and market its games in a manner similar to the way books and film are marketed—according to subject matter and genre. Thus there are sports games, science fiction games, fantasy games, war games, and action adventure games.[1] Such categories make sense from a marketing point of view because consumers are used to thinking about entertainment in terms of genre. Even so, these categories often reveal little about the basic design approach or the fundamental nature of the gaming experience. Whether played on the computer or elsewhere, games rely on a set of common skills.

GAME SKILLS

All games rely on three different skill sets in various proportions: physical skills, intellectual skills, and role-playing skills.

The physical skills required in traditional games may be overtly athletic in nature, such as those needed in sports like track, baseball, and football. A game can still be primarily physical in nature, though, without involving either great strength or endurance. This is true of such games as darts, bowling, jacks, and pinball, all of which involve more dexterity than muscle. Such games have their mental requirements as well: the ability to concentrate and to coordinate mind and body. By and large, though, success is determined by the participant's ability to complete some physical task.

These games have their counterpart on the computer screen. They involve not strength but dexterity and quick reactions, as in the notorious "twitch games," also known as arcade games. The key skill required is the ability to manipulate the cursor in order to evade and destroy on-screen opponents.

A second type of game involves more cerebral skills, including the ability to calculate and analyze. Traditionally, these are parlor games involving the use of cards, dice, boards, and score pads. Success in such games often requires a combination of strategy and luck. In cards, no amount of skill can help a player who continually draws losing hands. The same can be said for board games like Monopoly, in which a player's strategies for accumulating inventory may be foiled at every turn by the roll of the dice. The purest sort of strategy games, such as checkers and chess, minimize the effects of chance by eliminating luck determinants like dice, and instead pit opponents one against the other. All these types of strategy games have their counterparts on the computer.

A third and final type of game involves fantasy role playing. These games are characterized by the fact that participants act out a part and then stay within character in that role. Children, of course, engage in this type of game all the time, whether playing house or war, whenever they place themselves in fantasy environments. A more structured version of such a game might be *Dungeons & Dragons*, which at its most ornate is a strategy game involving costumes and role playing by the participants. Other adult versions include pantomime games and parlor mysteries in which party goers play the roles of suspects at a murder scene. Such role-playing computer games are often referred to as *adventure games*.

ELEMENTS OF GAMES

Games, no matter what type, have certain things in common. The most obvious commonality is that almost all involve competition, whether it is one human being against another, a human being against himself or herself, or a human being against a machine.

Because competition is integral to the vast majority of games, most games also involve some way of determining a winner, usually by some means of score keeping. Accumulation of the most points is one way of claiming victory, but there are other ways of vanquishing opponents: by taking all their territory, by cornering them, by outdistancing them, or by destroying them.

Although some games require nothing but our imaginations, most require some sorts of specialized materials, such as balls, bats, and playing fields. These extra materials take three forms. First are the basic tools or *implements* needed to play the game. In football, for example, one tool is required above all others, and that is the football itself. The second primary element in most games is some sort of defined space, or *territory*, whether that be a playing field or a checkerboard. Together the tools and the territory form the parameters of the game, defining its physical space and the means of locomotion through that space. In Monopoly, for example, the physical space is the game board, and the dice are the tools for determining movement in that space. In football, the physical space is the playing field, with its yard lines, end zones, and goal posts.

The third primary element of game playing experience is the *inventory*—the stuff that a player accumulates or loses during the course of the game. Inventory differs from the implements in that the quantity of inventory is variable, and the player manipulates that inventory for strategic advantage, accumulating it or expending it according to the game situation. To return to the game of Monopoly, for example, money and property are inventory, as are hotels and houses to be placed on that property. Success in the game depends on the manipulation of that inventory. The inventory is quite distinct from the dice or the cards in the community chest, which rather than being in the player's control are simply the implements by which the game is played.

The difference between implements and inventory, though, is not always clear-cut. In a game like checkers, the inventory and the implements are one and the same. Players use the individual checkers to get around the board and also maneuver them for strategic value, attempting to diminish their adversary's inventory while protecting their own.

As we have seen so often, categories tend to blur or overlap, and the same is true for games. Even so, we can make some general comments about games, and these observations hold true for computer games as well.

Most games on the computer involve a physical space, or at least the graphic rendering of a physical space (see Figure 6.1). The most common game implement is the cursor itself, though its appearance and function may vary greatly. In computer games, as in traditional games, successful play often involves manipulation of game inventory in some kind of contained environment, and the most brilliant strategies are sometimes foiled by bad luck. Because good games are typically competitive in nature, a good computer game needs to provide that competition—either through graphic renderings of an opponent, through scoring mechanisms that pit the player against the machine, or by providing outcomes and conclusions that are emotionally satisfying.

Another aspect of all games are the rules. Some rules govern movement in the physical space, whereas others govern the accumulation and use of inventory. In traditional games, players rely on written rules for the adjudication of disputes, and interpretations of those rules can vary greatly. On the computer, the rules are often more absolute. They are programmed into the software, and violations are often either impossible or involve severe penalties in game play.

Figure 6.1
A Computergraphic
Rendering of a
Physical Space.
Pictured is an
adaptation of pinball,
the classic arcade
game, to the home
computer. (*Crystal
Caliburn,* Starplay
Productions.)

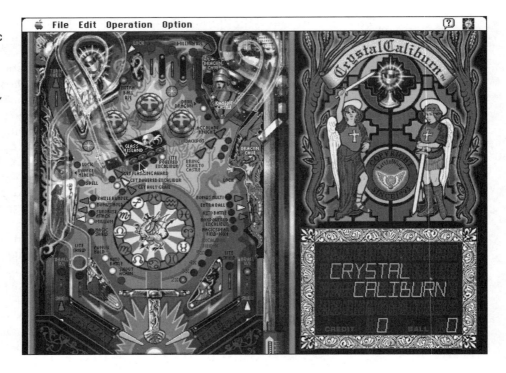

Types of Computer Games

In the following section we will examine three major types of computer games: arcade games, strategy games, and adventure games. We will also consider the special features of on-line games.

ARCADE GAMES

Arcade games have their roots in the old traveling fairs, in the shadowy tents and sideshows of circuses and carnivals. They are games of physical skill and dexterity, such as throwing balls at milk bottles and shooting floating ducks. They present challenges that are not as easy to overcome as they might seem, as the balls are lopsided, the milk bottles are bottom-heavy, and the ducks— well, they are not true targets at all, but images on a mirror. These old midway games are the ancestors of those now played in the enclosed arcades, those small dark rooms at the end of the boardwalk where men and women, girls and boys, all hunched over pinball machines and other whirring devices, feed in coin after coin, oblivious to the outside world. Like the outdoor games at the fair, these games have their built-in impossibilities, their tricks and their illusions. The players know this, but they play on anyway, obsessed and addicted.

Electricity has long been an element of the arcade, contributing flashing lights and ringing bells, or "the juice and the holler." In the 1970s, though, a new element—the computer—entered the arcade. Just as the computer has changed much else in society, it has also changed the arcade.

One of the first computer games to invade the arcade was *Asteroids*, itself an adaptation of electronic "shoot 'em ups" that computer programmers played on large mainframe computers. Advances in integrated circuitry allowed the same game to be played on a miniature, specialized computer hooked up to a black-and-white monitor. Players used knobs and buttons to control the motion of a spaceship and to shoot phosphorescent streams at encroaching asteroids. It was a simple game, but people loved it. Before long there were imitators, who added color to the screen and, eventually, more sophisticated animation. Within a few years, people were able to play these games at home using miniature computers hooked up to their television sets.

Pinball games, target shooting, miniature golf, mechanical baseball, foozball, race-car simulations, maze games—almost all the old standbys of the boardwalk arcade have been digitized and adapted to the home computer. Just as successful play in these older versions involved certain types of manual dexterity, success now involves the ability to manipulate a cursor on a computer screen. This manipulation takes two forms: evasion and attack.

The competition, as in old arcade games, is against the machine. The implements are digitized pinball flippers, electronic baseballs, animated spaceships—all controlled from the computer keyboard (or with a joystick). In the old arcade, keeping track of inventory was simple; the player had three balls to throw, for example, then the game was over. Similarly, in computer arcade games the players are allotted a certain number of opportunities. They have three spaceships, for example, to navigate into the great beyond, and when all three have been destroyed the game is over.

Computer games do add a dimension lacking in the old arcade, and that is the ability to animate the player's adversary. In an old-style duck shoot, the player would shoot at cute yellow creatures floating helplessly by on a stream of moving water. In the computer version of this game, the same ducks are able not only to take evasive action, but also to pursue their assailants and fight back, to attack and destroy.

In addition to expanding the role of the adversary, computers can also maintain and even increase a player's interest by advancing the level of difficulty. When a player becomes accomplished enough to complete one game task, another task presents itself. The task at this new, higher level of play is sometimes simply a more difficult variation of the task just completed; other times it is a new task altogether.

Consider the evolution of the maze game. In the old arcade, such mazes were simple wooden boxes with a lever-controlled tiltable playing surface consisting of a narrow maze into which holes were drilled. The player's task in such games was to tilt the table in such a way as to guide a small metal ball past the holes to the other end of the maze. Later, a very successful electronic game, known as *PacMan*, took this same concept and amplified it using the computer. The balls were now two-dimensional bright-yellow cartoon characters rendered on a screen, but the concept was the same. The player's task was still to guide these characters through a maze, only now they gobbled up pellets as they went, all the while under pursuit by enemies. The player's goal was to avoid the enemies and consume all the pellets. When this was accomplished, the player advanced to a higher level of play, to a new screen with different background colors but the same task, except that everything happened faster.

Maze games have continued to evolve. With new animation techniques come more complex on-screen figures. Instead of manipulating balls through a maze, players manipulate cartoon figures through jungles and caves. Users can make the figures run and jump, roll to the side, climb a ladder, fire a gun. The enemies, too, are more complex. When developing such games, designers make up story scenarios for the main character, giving him or her different enemies at each level of play, changing the setting and environment. Underneath all the animation and technical glitz, however, the principle of such a game is very simple: It is still a maze game, with its roots in the old arcade.

STRATEGY GAMES

Strategy games emphasize thought and reflection as opposed to dexterity, contemplation as opposed to physical action. They are more cerebral in nature and often engage the player in puzzles—puzzles regarding the manipulation of objects in space and time, puzzles about numbers and words, or puzzles that require the shuffling and reorganization of information. Sometimes, strategy games involve an element of chance in the form of dice or cards, and they might also require boards and a complex inventory for manipulation by the players. Often there are rules that take time to master. They include parlor games such as Hearts or Bridge or Canasta; board games such as Monopoly and Risk, or checkers and chess; and three-dimensional puzzles such as Rubik's Cube.

When computer game companies began their invasion of the home market, one of the first things they did was adapt these traditional games to the new medium. For games like blackjack and checkers, this has been a relatively easy task. The computer, after all, is an excellent device for making calculations and storing information, and programmers developed algorithms that allowed players to vary the intensity of the competition. Even so, most of these adaptations have not had the appeal of the original games because by their nature they require a human opponent. Much of the charm and interest and joy seems lost when playing a machine.

More interesting, perhaps, are outgrowths of these older types of strategy games developed specifically for the computer. One such outgrowth are the *motion puzzles*, variations of the slide puzzles children play with on long automobile trips. In the physical world, such puzzles consist of a hand-held tray filled with slatted and grooved tiles, and the user's task is to arrange the tiles into a predetermined pattern without lifting them from the tray. On the computer there are hundreds of variations of this simple game involving pixelated tiles and hidden obstacles. Often the goal is to remove the tiles, one by one, while avoiding booby traps that explode when the tiles are removed. Other times the task is the arrangement of moving tiles into patterns, using the mouse to snatch them as they drift down from the top of the screen, then rotating and dragging them into some sort of receptacle. One of the most well-known games of this sort is *Tetris* (see Figure 6.2).

Figure 6.2
A Motion Puzzle. In this classic motion puzzle, the user tries to manipulate the falling tiles until the entire space is filled, with no empty (black) areas in between. In this uncompleted game, the player has made several miscalculations in the bottom three lines. (*Tetris Max,* Steve Chamberlain.)

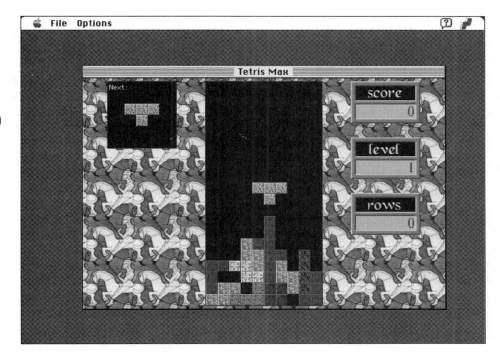

Other strategy games retain many of the elements of board games, with the computer taking over the randomizing function of the dice as well as tracking functions. This is particularly true for traditional sports-strategy games, which typically involved not only dice and board, but also game pieces whose performances were based on statistical analysis. A typical play sequence might involve setting up the pieces on the game board, then rolling the dice, and finally making mathematical calculations with paper and pen in order to determine the results of the play. The computer has made much of this instantaneous.

A complex form of computer strategy game that attracts a large and stable audience is the war game. Like their board counterparts, these games involve movement of pieces on a field of action, and that movement is tied to mathematical calculation. Pieces are assigned numerical values according to their relative strengths, which vary according to such factors as weaponry, available ammunition, supply lines, and number of casualties. The player in effect takes the role of supreme commander, and the battlefield variables are presented to him or her for analysis and interpretation. War games also feature some narrative context, in that the battles are often reenactments of famous military engagements. The programs may even take into account the weather on the actual day of battle, the personalities of generals, the terrain, and other historical factors. In this sense they use some of the traditional vehicles of storytelling to recreate a dramatic environment for the player.

Another form of strategy game are the world-building games, such as the *Sim-City* games from Maxis (see Figure 6.3). In the this type of game, the player takes on the role of world builder—managing, for example, the growth of a city. The player starts with an empty landscape and some inventory with which the player builds a town with residential, industrial, and agricultural sections. As the town gets bigger, the needs of the citizenry grow more complex. The player's job is to balance the various needs of the citizenry and keep the world functioning. As in other strategy games, the computer keeps track of the inventory and does the essential calculations. It also provides occasional random disasters—a fire, an earthquake, a flood—that send these managed cities spinning toward destruction.

ADVENTURE GAMES

Adventure games can be distinguished from other games in that they require a degree of role playing by participants. These games often require stronger, more coherent story lines than other types of games, and they also involve character development. The usual dramatic structure centers on a quest in which the player is sent on a mission into an imaginary realm. Often the player is accompanied by other, computer-generated characters who have their own personalities and can help or hinder the quest according to their skills and temperaments. The joint goal may to be slay a dragon, or to remove some ancient curse, or to solve a murder mystery. Adventure games often incorporate elements of the classic computer strategy game and sometimes require certain arcade skills as well. The player

Figure 6.3
A Strategy Game. The player's task here is to keep the company functioning, supplying food, power, and shelter for the citizens while maintaining environmental and economic balance—all in the face of various natural and social disasters. (*SimCity 2000®*, Maxis® © 1994. All rights reserved.)

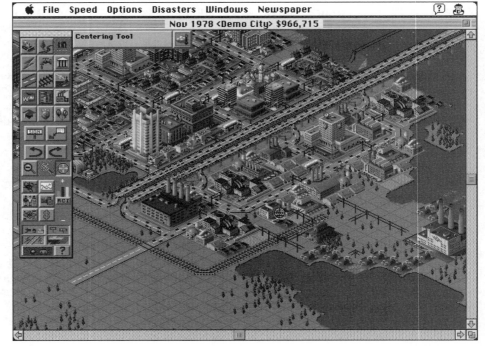

seeking entrance to a castle, for instance, may need first to solve a word puzzle, thus unlocking the castle gate. Once inside, the player may then have to engage in an arcade-style battle to defeat a dragon.

Role-playing games have their roots in a text-based game called *Adventure*, first played by computer researchers on mainframe computers. The original game employed no graphics, only text descriptions of an imaginary underground realm. It was the player's task to proceed through that world, learning its geography, accumulating weapons and treasure, and battling dwarfs and other demons. The player accomplished this by typing commands such as "Go south" and "Pick up knife."

In a text-based game of this sort, players might find themselves confronted with the following description:

> You are standing next to a hollow tree in the middle of the forest. Inside the tree is both a bottle that looks like it might hold a magic potion and a beehive dripping with delicious honey. A path leads away from the tree in both directions. To the east a small stream is visible; to the west you see nothing but trees.

Interaction with this particular scene might go as follows, with the upper phrases (all capital letters) being user commands, and the lower phrases being the computer's responses:

> TAKE THE POTION
Taken.

> TAKE THE HONEY
A flock of bees swarms out of the hive. They are angry.

> KILL THE BEES WITH POTION
Nothing happens.
Now the bees are very angry.

>RUN
Which direction?

>RUN WEST
Too late. The bees have stung you. The bee venom is poisonous, and you don't have long to live.

>USE POTION
The potion bottle is empty.

>RUN
You are too weak to run.

>CALL DOCTOR!
I don't see a doctor. Besides, it's too late.
****You have died.****

With the advent of digitized graphics and sounds, game developers elaborated on the structure of these early adventures games, rendering the environment with 3-D graphics and replacing the text commands with visual navigation controls. *Return to Zork*, for example, is a direct descendent of such games, only in a CD-ROM format. In this game, the player embarks on an adventure to the idyllic

Valley of the Sparrow, only to find that the valley has been desecrated and a nearby town destroyed. As in the old text-style games, the player still must navigate the landscape, though now the inventory items are graphically displayed, as are the enemies. Also, the player has a chance to engage in seemingly open-ended conversations with on-screen characters in the form of actors whose images have been recorded on video (using a *blue-screen* technique) and superimposed on the landscape.

As adventure games have continued to evolve, a trend has emerged among game developers to regard storytelling and character development as something more than an enhancement to game play—as legitimate ends in themselves. This has given birth to a new genre of games, sometimes called *interactive cinema*, which attempts to effect a fusion of storytelling and traditional gaming techniques. Examples include *Seventh Guest*, *The Journey Man Project*, *The Daedalus Encounter*, and *The Psychic Detective*—as well as *Myst* and *In the 1st Degree*—all of which are role-playing games that attempt to integrate story and character into an interactive structure. Some of these games have attracted a loyal audience, as well as considerable enthusiasm among game reviewers. With few exceptions, however, by and large they have been received somewhat coldly by hard-core gaming enthusiasts, particularly those who prefer more traditional, arcade-style challenges.

ON-LINE GAMING

Games have been available on the Internet for quite some time. Many of these have been text-based games, similar to *Adventure*, in which the user navigates through an imaginary, unseen world using text prompts and verbal commands. Others have been on-line renditions of chess and checkers and various *Tetris*-style games that have used simple bit-mapped graphics to render their visual dimensions. Until recently, these renderings have been fairly crude.

One of the features of on-line gaming is that it opens the possibility of multi-player games. In other words, it makes it possible for one person in Wisconsin and another in Shanghai to play a computer game against each other. For the most part, such games have been limited to chess, verbal fantasy games, and other games requiring only simple graphic interfaces. More recently, though, Java and Shockwave software has made it possible for remote players to shoot it out on the Net in simple twitch games, though these are below the arcade quality available in most home software.

Another approach to multiplayer gaming has involved dedicated on-line services. In this approach, it is necessary for each user to have a software version of the same game, in either disk or cartridge format, depending on the platform. The players load the software into their computers, just as normally might be done; then they use modems to dial up the on-line gaming service. The service's computer connects two players to the same game, and they battle it out. This is made possible by the fact that the central computer reads the mouse movements of the first player, then manifests those movements on the second player's screen using graphics that are stored in the second player's local software. Because of bandwidth problems, it is difficult to transmit graphics stored on remote servers, unless the players are operating on closed intranet systems or otherwise have access to greater bandwidth than that provided by conventional phone wires. (For further discussion of bandwidth, see Chapter 9.)

Another approach to on-line gaming has its roots in early experiments at Lucas Film in a project known as the Lucas Film Habitat.[2] In that project, players assumed fictional identities and interacted in an imaginary computer world. Their fictional personas, known as *avatars*, interacted with other avatars and created a virtual community with social rules, marketplaces, political campaigns, and the like. The avatars were able to move about and speak with one another through cartoon-like text bubbles containing words, typed in by the players. This approach incorporates some of the techniques of traditional computer adventure games, such as inventory, guide figures, and navigational tools. In order to accommodate a large number of players, though, such a system requires a fairly powerful server and rapid modems, as well as sophisticated transmission codes capable of activating graphics stored in a user's CD-ROM in response to the movements of other users in remote locations.

Recently, Fujitsu and a rival firm, Knowledge Adventure World, conducted simultaneous experiments in this field. In one prototype version, Alice-in-Wonderland avatars wandered down surreal hallways, many of which were decorated with billboards and kiosks that, when suddenly activated by the presence of the user, began playing advertisements and flashing commercial messages. The end goal was to create a giant world's fair, including pavilions grouped according to subject matter, where users—under the guise of their avatars—could explore, chat, witness sporting events, play games, or conduct business.[3]

Figure 6.4
An On-Line Drama. This popular interactive soap opera, *The Spot,* relies primarily on text and photos, but also supplies some audio as well as video clips that can be downloaded and viewed off-line. The style is episodic, and events are related from alternating points of view. (*The Spot,* http://www.thespot.com.)

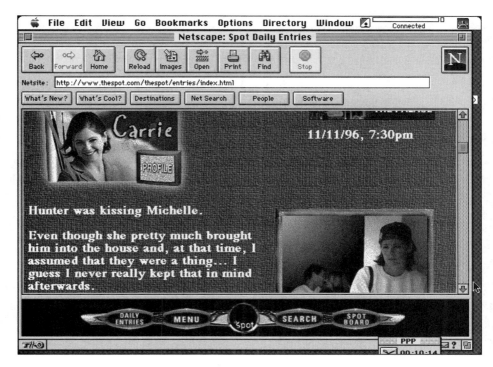

Another approach to on-line gaming occurs on the *World Wide Web*, and it is less related to traditional computer games than to interactive cinema. These interactive shows operate according to the model of serial television, with new installments weekly. They also borrow from television genres, pitching themselves as soap operas, or mysteries, or sci-fi adventures, or contemporary dramas. One early experiment in this format, *The Spot* (see Figure 6.4), was a soap opera tracing the exploits of Gen X-ers in a Santa Monica beach house. Like many of its imitators, *The Spot* in its early days was largely text-based, with hot spots providing graphic annotations. Each episode was told from the point of view of a different character, and links to previous episodes were used to describe the same or related incidents from other points of view. It was also possible to download video clips. As the real-time video and audio capabilities on the Web have increased, designers have juiced up the hot spots by providing instantaneous video and audio.

Other interactive dramas on the Web, such as *Madeline's Mind*, have also taken advantage of the increasing media capabilities, building virtual environments that are based on a more explorational model. Instead of reading text, the audience probes a visual environment, clicking on objects to uncover the storyline. In this model, users delve into a new screen or realm each week, and the story moves incrementally as they experience these new screens. Although these experiments in interactive narrative are still limited by bandwidth considerations, the media capabilities of the Web are constantly expanding. As they do so, so too will the designer's conception of form, as well as the audience's expectations of this new format.

The Game-Scripting Process

The writing and design process for computer games follows the same general design process outlined in Chapters 2 and 3. This process begins with concept development, in which the initial idea is explored in terms of content and its interactive potential. Concept development is followed by a treatment phase, in which the various game elements are expanded and integrated. Because game development can be so complex—involving the charting of story and character through shifting interactive contexts—the treatment stage itself is often a two-part process, involving first the production of a preliminary treatment and then a full or expanded treatment. After the treatment phase is completed, a full production script is written, in which the combination of audio, visual, and textual elements are compiled on a screen-by-screen basis.

DEVELOPING A GAME CONCEPT

The earliest computer games were designed by solitary programmers working in their free time without financial backing. Because this occurred before digitized sound and video, the focus was by necessity on the gaming experience itself. Under these circumstances, an individual programmer could conceive, develop, and program a game almost entirely alone. In the past twenty years, though, computer gaming has become big business. The development process has become more complex and now involves the skills of many individuals. Design teams have grown, and programmers do not necessarily control the creative process. Large firms such as Electronic Arts now develop games using a Hollywood production model employing producers, graphic artists, writers, and programmers. Concept development might be the responsibility of any one of these people or of several people working in concert.

In Chapters 2 and 3, we learned that good concept development in interactive media must consider not only the content but also the nature of interactivity—how the user will experience the program, what sorts of actions the user can take, and what the results of those actions will be. The same general conceptual concerns underlie game development. In designing games, however, one of the first things a designer must consider is the genre, or type of game being designed. Different genres have different audiences, and these audiences bring different expectations to the screen.

In arcade-style games, players seek a fast-moving visceral experience, one that immerses them in a clear task that requires immediate and impulsive action. The pleasure of a twitch game lies in the deftness of movement itself, in the joy of evading one's enemies—and in the even greater joy of destroying them.

The challenges to players in such games are almost always challenges of dexterity regarding their ability to maneuver the cursor. This is a completely different challenge than that presented players in a strategy game, an adventure game, or in interactive cinema. It implies different things about both the design metaphor and the level of character and story development required or desired, by players.

Flashback, for example (see Figure 6.5), is an arcade-style game that utilizes an episodic narrative and plot, yet at its roots it remains a shoot-and-jump arcade game involving the manipulation of an on-screen character past various enemies and dangers. As in other similar games, players shoot and destroy enemies, working their way through levels of increasing difficulty until finally all enemies are destroyed. Players' concerns in such a game revolve around manipulating the cursor, the characters' different possible evasive abilities, and the kinds of weaponry available. Of course, exciting enemies and vivid settings are important too.

In *Flashback*, the primary on-screen character is a graduate student and part-time secret agent, Conrad Hart, who suffers from amnesia as the result of a previous tussle with the bad guys. His mission is to regain his memory and destroy the aliens who threaten humanity. Although the underlying filmic story is revealed in bits and pieces as a reward for success in battle, in actuality the focus in the game is on arcade skills. The main drama lies not in the story, but in players' failure or success at manipulating the objects on screen. The more successful, the farther a player advances and the more plot points are revealed.

At some point during the development of a game like *Flashback*—or of any game, for that matter—the designers must define their concept. It must be clear, and it must define the game in terms of genre, narrative stance, and players' goals. It is best if the game premise can be boiled down to a few concise sentences. For *Flashback*, the designers' premise might have read as follows:

Figure 6.5
A Shoot 'Em Up.
Though this game,
Flashback, uses
character and story
elements as
background, it is in
many ways a classic
twitch game involving
the ability to control
the on-screen
movement of figures
using a keyboard or
joystick. (*Flashback,*
MacPlay.)

Our goal is to create an arcade-style shoot'em up with an underlying narrative, a game in which the player takes on the persona of an amnesiac secret agent. The plot elements will be revealed in the course of combat, as the player battles an alien race out to destroy humanity. The player's goal is to destroy that alien race.

CREATING A CONCEPT TREATMENT

Once a premise has been established, it is common practice among many developers to commit that premise to paper as part of a concept document, in effect an early-stage treatment of the material. Length and exact content vary according to the needs of the producer, but the major elements (see Figure 6.6) are predictable: a statement of the game's premise; a summary of the story elements, including plot, characters, and setting; and a summary of the primary gaming elements. The concept document might also touch on such matters as design metaphor, interface design, and the like.

The path a game takes from concept to script is not always the same. In a simple arcade game, a traditional script may never be written because many arcade games involve little if any content development; the focus instead is on image and movement, and so the creative process revolves around the interaction between graphic artists and computer programmers. The development process in such instances might go directly from concept to production.

Figure 6.6

Elements of a Game Treatment. The first three major sections (I, II, and III) of this treatment outline—Game Premise, Story Overview, and Character Bible—all deal primarily with content. Section IV, Game Play, deals primarily with questions of interactive structure. Section V, Media Elements, is concerned with practical matters of implementation as well as the look and the feel of the program. Though treatments may differ considerably in their use of headings and their organizational structure, most will deal with the fundamental elements listed here.

I. GAME PREMISE
Describes the main premise, the player's role, the major tasks and levels, and the type of game.

II. STORY OVERVIEW
Delineates the narrative action and major events of the game.

III. CHARACTER BIBLE
Whereas some treatments may contain only a character list, others often contain full bibles, with detailed descriptions of each character and their personal histories, motivations, and roles in the game.

IV. GAME PLAY
    A. Design Metaphor
    B. Game Methodology
    C. Interface Design
        1. Means of Navigation
        2. Task Considerations
           (Interviews, inventory accumulation, etc.)

V. MEDIA ELEMENTS
    A. Platform and Format
    B. Visual Look and Feel

This is not always the case in other types of games. Because strategy games are often quite concerned with the manipulation of content, they sometimes require a great deal of research that is eventually compiled into a document sometimes called a *treatment*. (Other times, as mentioned earlier, it is called a concept document.) The treatment for a war game, for example, might involve a basic narrative scenario communicating the historical situation and the personalities of the combatants, as well as mitigating factors and other dramatic elements the designers deem significant. The same sort of considerations come into play for world-building games. Designers need to define the basic social dynamics of the world being created, and content documents can help them in that task. Such documents might also sketch out scenarios in which probable game experiences are explored and perhaps refined.

Written treatments and scripts are most widely used in adventure-style games, particularly in role-playing adventures that use techniques from interactive cinema. To get a better idea of treatment elements, consider the following concept treatment for a proposed mystery adventure, *The Moonstone*, based on the Wilkie Collins novel of the same name. This preliminary treatment assumes that the designer has the luxury of a publisher with a large enough budget and the technical abilities to enable use of both 3-D rendering and blue-screen video.

PRELIMINARY TREATMENT FOR *The Moonstone*

I.   GAME PREMISE

In this game, the user plays the part of the legendary Sergeant Cuff, a Scotland Yard detective who must trace down the whereabouts of an ancient diamond known as the Moonstone. This is primarily a role-playing game in which success depends on strategic manipulation of inventory. The player must interrogate witnesses and assemble evidence while navigating a three-dimensional environment. The game takes place in three stages, each analogous to a single day of the investigation.

II.   STORY OVERVIEW

The story begins on Rachel Verinder's twenty-first birthday, when the Moonstone is given to her as a behest from a recently deceased uncle of dubious repute. Unknown to Rachel and the others at the party, the Moonstone carries a curse—and the jewel is stolen from her the night of the party.

At this point, Sergeant Cuff is called into the investigation. It is his job to search the family estate in the English countryside, interview the participants, and recover the Moonstone. In the early parts of the story, his suspicions are directed toward the servants, in particular toward a scullery maid who has spent time in a woman's reformatory. As the story progresses, however, Cuff discovers that the answer to the mystery lies in the heated rivalry between two of Rachel's suitors, both of whom attended the party.

Each of the three days of the investigation closes on a dramatic event. The first day, for example, ends with the unexpected death of the prime suspect, the scullery maid. The second ends with the sudden disappearance of Franklin Blake, Rachel's rejected suitor and childhood sweetheart, and the third day ends with a revelation implicating Rachel herself in the crime.

III.   CHARACTER LIST

Sergeant Cuff—the point-of-view character. He is a hard-bitten, Scotland Yard detective with a fondness for rose gardens.

Rachel Verinder—an attractive young woman of an old family in the English countryside. She is high-strung and emotional.

Madame Verinder—Rachel's mother, who is protective of her daughter.

Franklin Blake—Rachel's childhood sweetheart, with whom she has had a romantic attachment for years, though nothing ever seems to come of that attachment. He is a vagabond of artistic nature who is encumbered by gambling debts.

Geoffrey Ablewhite—Rachel's other suitor, apparently selfless and wealthy and of good background. He detests Franklin Blake.

Gabriel Betteredge—the family servant who serves as the guide figure in the game. He is helpful to Sergeant Cuff, but only to a degree for he is tight-lipped with any information that might lead the investigation in directions injurious to his employers' reputations.

The Scullery Maid—a former prostitute and thief who now works in Verinder home.

IV. GAME PLAY

DESIGN METAPHOR: The investigation takes place in the house and grounds of the estate. As such, the process of the investigation itself—with its movement from place to place—provides the underlying structure or design metaphor. Among the places that can be visited are the parlor, the servants' quarters, the bedrooms, the dining hall, the kitchen, and various sitting rooms and porches. Exterior locations include the butler's cottage and the lagoon.

GAME METHODOLOGY: The task here is largely an exploratory one, in which the player explores the various locations on the estate, taking items into inventory for later use, gathering clues, and interrogating witnesses. These tasks are interrelated. The player must conduct successful interrogations in order to find clues, but the player must also have certain evidence in hand in order to gain the cooperation of the witnesses. This is complicated by the fact that the player can hold only so much evidence in inventory.

In addition, a time factor defines game play and adds suspense. Certain tasks must be performed during each of the three days if the investigation is to succeed, but the investigation moves forward whether they are accomplished or not. Each day is punctuated by a number of dramatic events, some of which are fixed, others of which occur as a result of the player's actions.

INTERFACE DESIGN: The house and grounds are rendered in 3-D graphics, and the player moves about by pointing and clicking in the direction he or she wants to go. In the beginning, the player can move from place to place only by traveling through adjacent rooms, but as play continues the player will discover a network of secret passages that allows for movement between areas on the estate that are not adjacent. The interface must also be structured in such a way as to accommodate the accumulation of inventory and to provide a means for interviewing characters as they are encountered throughout the estate.

V. MEDIA ELEMENTS

FORMAT AND PLATFORM: This program will be released as a double-disk CD-ROM playable on the current generation of IBMs, with Pentium processors and 16 Megs of RAM. Comparable versions will be released for Power Macs.

VISUAL LOOK AND FEEL: The program will employ approximately two hours of blue-screen video, to be rendered against a three-dimensional environment recreating the rooms of an old, somewhat austere house in the English countryside. Though a great deal of attention will be paid to realistic detail, the overall intention is not a realistic rendering of Victorian times. Rather, the intent is to use compositing and retouching techniques to create a visual world that seems both dark and lush with possibility, full of repressed sensual impulses and desires that manifest themselves, however briefly, as the player investigates first the theft of the Moonstone, then the mysterious death of the scullery maid.

Keep in mind that this is not a full treatment, but only a preliminary treatment or concept document. A full treatment would take this basic outline and flesh out the various elements much more completely.

Before that could happen, however, the designers would have to make some decisions about the interface. The interface described here is a very challenging one, incorporating a three-dimensional navigational system as well as tools for interviewing and gathering inventory. Basically, two approaches can be taken: The tools can be on-screen at all times, or they can be hidden and appear only when they are needed. Either way has its advantages and its drawbacks, but making that decision is only part of the problem. Designing a mechanism and approach for interviewing witnesses is in itself a challenging task, and the designers must make certain fundamental decisions about how the interview process will take place. The approach the designers took in *In the 1st Degree* is depicted in Figure 6.7. Such decisions are important, because they affect the structure of the story and of the game itself.

One way designers might begin this process would be drawing a flowchart of possible routes of movement through the house and grounds, as well as sketching out a sample room and imagining the different activities that can take place

Figure 6.7
Interface for Interrogations in *In the 1st Degree*. During the interrogation of witnesses in this program, players first choose a topic (at left), then questions from a list (right). After a question is answered, the player may choose another question, change topics, or pursue other witnesses, tapes, or documents. (*In the 1st Degree*, Broderbund.)

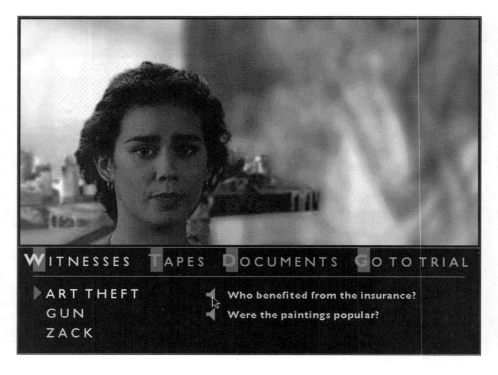

WITNESSES   TAPES   DOCUMENTS   GO TO TRIAL

▶ ART THEFT   ◀ Who benefited from the insurance?
  GUN   ◀ Were the paintings popular?
  ZACK

there. Once this has been done, the concept document would be expanded into a full treatment.

COMPOSING THE FULL TREATMENT

The full treatment explains the story in greater depth and more precisely defines the events and goals of each level of play. It might also include a much more elaborate character bible, describing the characters and their motivations in detail. Moreover, it would list and describe the various items of inventory, including their relative importance and function. (For further descriptions of treatment elements, see Chapter 2 and Figure 2.5.)

The full treatment, in other words, develops everything that is implicit in the concept document. It answers the unanswered questions about methodology and story that were raised by the concept itself: Who stole the Moonstone and why? How was it stolen? What sequence of events must the player initiate in order to successfully identify the criminal?

The full treatment, then, must accomplish two tasks: It must define the underlying story in all its specifics, and it must tell how the user is to solve the problem that the game presents. A concept document does not necessarily do either of these things but instead merely presents the game in an open-ended fashion and suggests how it might be played.

Once a full treatment has been written, the scripting itself may begin.

WRITING THE SCRIPT

One of the questions that always comes up in regard to scripting is format. What should a script look like? How should it be laid out on the page? The answers to such questions most often lie in the material itself. If a script is location-based, it needs to describe the primary elements of the setting, as well as what the player might see and do while in that location. If it is dialogue-based, the script might look much more like a radio play, containing mostly dialogue and little physical description.

In the following script, which shows what a portion of the production script from *The Moonstone* might look like, the scene is the parlor on the first day of the investigation. The script includes a short video clip as an obligatory cut, after which the player has two choices: pursue the two witnesses, each of whom flees in a different direction, or stay and examine the evidence in the room. If the player stays, the investigation will be interrupted by an angry Madame Verinder.

PRODUCTION SCRIPT FOR *The Moonstone*

SCREEN 5.0

THE PARLOR—DAY ONE—FIRST VISIT
A parlor in the country style, a simple, comfortable room for the receiving of guests. There is a fireplace, a bookshelf, a love seat and table, and a chair. Significant objects include a stack of letters resting on the love seat; a portrait album and family history on the bookshelf, and also a black ledger containing financial statements for the estate.

There are three entries to this room: an open entry from the main hall; a side door, hidden from view, that leads to a sitting room; and an open door leading to the porch.

SCENE 1

OBLIGATORY CUT—(FIRST VISIT ONLY). The young mistress of the house, Rachel Verinder, stands in heated conversation with her sweetheart, Franklin Blake. Neither seems aware of Cuff's presence.

RACHEL: How could you?

FRANKLIN: I haven't done a thing. My feelings for you...

RACHEL: Don't lie to me.

FRANKLIN: Rachel, my dearest...

She slaps him. Then, looking up, she notices Cuff.

RACHEL: What are you doing here? Get out of this room, or I will have you forcibly removed from this house!

Rachel stomps off, leaving by the side door. Blake gives Cuff a soulful look, then disappears onto the porch.

END OBLIGATORY CUT

HOT SPOTS (ROLLOVER)
PHS1—PORTRAIT ALBUM
PHS2—LETTERS
PHS3—BLACK LEDGER

EXITS
PX1—ENTRY TO MAIN HALL
PX2—DOOR TO SITTING ROOM (locked)
PX3—OPEN DOOR TO PORCH

TIMED VISIT
If player stays in this room rather than exiting, Madame Verinder appears. (see MVC1)

HANDLING NONSEQUENTIAL ITEMS

A script for a program such as *The Moonstone* might consist of many segments similar to the one just presented, one for each visit to each room in the house. But because these segments are not always experienced by players in a predictable order, some decisions need to be made regarding how to organize the script. In this case, it might be organized according to location, with events occurring in a particular place grouped together in the script and numbered accordingly. The designations PHS1, PHS2, and PHS3 in the script constitute a numbering system for the hot spots—Parlor Hot Spot 1, Parlor Hot Spot 2, and Parlor Hot Spot 3—and gives the reader a description of the material contained in each hot spot. Similarly, PX designations refer to parlor exits and MVC1 refers to Moonstone Video Clip 1. The logic behind such numbering systems is situational, but almost all interactive scripts contain some kind of indexing system. This is one of the primary ways in which writing for interactive media differs from traditional scripting.

Another formatting challenge concerns the scripting of the various dialogue exchanges between characters in a given situation. How, for example, would a writer format all the possible exchanges between Sergeant Cuff and Franklin Blake if Cuff followed Blake onto the porch and engaged him in an interactive conversation? In such circumstances—when a player is given a multiplicity of choices—it is often all but impossible for the writer to render all of the possible dialogue in a strictly sequential fashion on the page. The writer then must rely on an indexing system that relates bits of written dialogue to a flowchart in order to recreate the movement of conversation as experienced by the player. Figure 6.8 shows a simplified example of how such a dialogue exchange might be graphically depicted and coded to relate bits of dialogue to the flowchart. A more complex example is presented in Appendix A regarding the charting of dialogue for *In the 1st Degree*.

Figure 6.8
A Method for Charting Multilinear Dialogue. In scripting multilinear dialogue, writers/designers must coordinate flowcharts (a) and dialogue segments (b) to track the flow of conversation in shifting contexts. The flowchart is often drawn before any dialogue is written. In this flowchart, rectangles contain courses of action presented on the interface to the player, who assumes the persona of Sergeant Cuff; ovals indicate Franklin Blake's responses. To identify the various dialogue streams, match the codes inside the rectangles and ovals with those in the dialogue segments (next page).

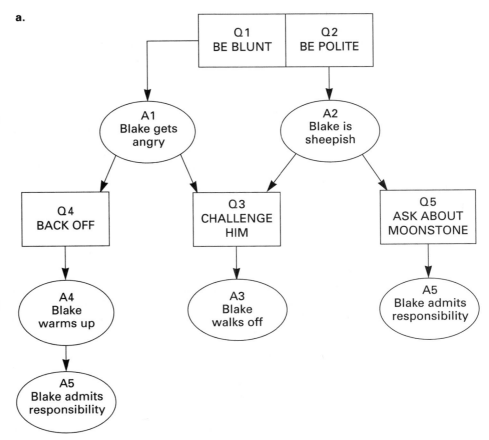

Figure 6.8 (continued)
Dialogue Segments.
Match these coded
segments to the coded
rectangles and ovals
in (a).

Q1: **BE BLUNT**
CUFF: What's going on between you and Miss Rachel, my friend?
That was quite an argument!

A1:
BLAKE: That's none of your business, Sergeant Cuff, no matter your
reputation.

Q2: **BE POLITE**
CUFF: It's pleasant here on the porch.

A2:
BLAKE: Yes, and I can use all the pleasantness I can find.

Q3: **CHALLENGE HIM**
CUFF: And how about your reputation? What is this lie Miss Rachel
speaks of?

A3:
BLAKE: There is no lie. (with finality) If you will excuse me, I must
be going to my room.

Q4: **BACK OFF**
CUFF: We all are a bit sullied, I guess, myself included. (pause)
Would you mind talking to me about this Moonstone.

A4:
BLAKE: I suppose I shouldn't mind—especially if it will clear my
name. And my conscience.

Q5: **ASK ABOUT MOONSTONE**
CUFF: How did this Moonstone come to be in the house?

A5:
BLAKE: My uncle, in his will, asked me to deliver it to Rachel on her
birthday.

b.

SUMMARY

There are four fundamental steps in the scripting process for computer games:
(1) developing the game's concept; (2) creating the concept document, or preliminary treatment, which outlines the fundamental narrative and design elements as they relate to the concept; (3) composing the full treatment, which is
based on an expanded understanding of the interface and develops and integrates
the narrative and gaming elements; and (4) writing the full script containing all
the visual descriptions, dialogue, on-screen text, and functionality parameters
necessary to bring the game into production. As we have repeatedly noted
throughout this text, there is a great deal of overlap between these steps and the
successful creation of a game script is a collaborative effort. The exact formatting of a game script varies from project to project, but almost all scripts require
some sort of indexing of nonsequential elements.

In the early days of game design, games were developed by programmers working alone, and written scripts were rarely used. This remains true for some arcade and strategy games, but with the continued cross-merging of gaming genres with cinematic techniques, the written script is becoming a standard rather than an exception.

E X E R C I S E S

1. What are the main elements of a traditional arcade game? A traditional card game? A board game? For each of these types of games, describe what is gained and what is lost when they are adapted to the computer.

2. In this chapter, the concept treatment for *The Moonstone* describes some basic interface specifications for this program. What are those specifications? Try to envision another interface with different specifications. Draw a picture of that interface. How would changing the interface affect the game?

3. Come up with a concept for an original computer arcade game. Write a preliminary treatment that outlines the basic game concept, the story elements, the game elements, and interface considerations.

4. Write a preliminary treatment for your own computer strategy game.

5. Write a preliminary treatment for an original adventure game.

E N D N O T E S

1. Bart Farkas and Christopher Breen, *The Macintosh Bible Guide to Games* (Berkeley, CA: Peachpit Press, 1996).
2. Chip Morningstar and F. Randall Farmer, "The Lessons of Lucas Film Habitat," in Michael Benedikt, ed., *Cyberspace: First Steps* (Cambridge, MA: The MIT Press, 273–302.
3. Domenic Stansberry, "Going Hybrid: The CD-ROM/On-line Connection," *New Media Magazine*, June 1995, 34–40.

7

Educational Multimedia: Scripting from an Instructional Design Model

Educational multimedia is a broad category. The term has been used to describe everything from how-to titles for the home, to specialized reference works, to computer-based training, to databases of text and graphics stored on remote sites on the Internet. In short, the term has been used to describe almost any kind of knowledge accessible via computer, particularly if that knowledge has graphic or audio components.

For our purposes here, we will not use the term to describe reference and informational databases, however ornate they may be. Rather, we will reserve it for multimedia programs whose primary purpose is to teach in an active and deliberate manner. These programs may entertain in the meantime; they may be funny, engaging, thrilling, or enthralling. The entertainment, though, is not an end in itself but a means toward the educational goal. The purpose of an educational program is to convey some skill or knowledge users need to enhance their abilities or their understanding. Theoretically, everything in the program should augment the educational purpose.

Developers of educational multimedia refer to three types of educational programs. First there are *foundational programs*, which teach skills and concepts fundamental to a given discipline. Because foundational programs teach basic principles, they often incorporate some way to measure students' level of achievement. Next there are *exploratory programs*, which encourage users to roam throughout the material. These programs, which are typically open-ended, teach students problem-solving and investigative skills. Finally, there are *management programs*, designed mostly for teachers to help in the daily management of the classroom. They include such features as lesson plans, portfolio management, and tools for tracking the progress of individual students. In some cases, a single program will incorporate all three program types, providing a management system for teachers while teaching both foundational and exploratory skills to students.

The Instructional Design Process

Regardless of what type of educational program is being produced or the subject matter taught, the fundamentals of the *instructional design* process remain the same. The process can be divided into four stages that are roughly analogous to the design stages first described in Chapter 2, with some special considerations to accommodate the educational context in which the programs are used.[1]

The first stage, the *proposal stage*, focuses on concept development, with special emphasis on educational needs and instructional goals. The second stage, *content and structural development*, concerns itself with formulating educational content, course plans, and interactive exercises, and then melding these elements into a treatment document. The third phase, *script design*, focuses on the production script. The fourth stage, *implementation and revision*, is somewhat unique to the instructional design process and emphasizes the continuing need for revision after a program has been implemented in the classroom. In educational programs, testing not only takes place before production, when prototypes are given trials and revised, but also continues after the final product is in use. Through testing, the designer discovers whether or not the program works as a teaching device and how it might be improved. If performance objectives are not being met, the problem likely is not with the students but with the program— and it is time to revise.

Theoretically, the writer/designer is involved with every stage of the instructional design process. In reality, however, the designer's influence often recedes as the production process begins. Implementation and testing are often done by specialists, and the writer/designer must rely on field reports and other data to determine how well the project is working and what kind of revisions and updates need to be made.

THE PROPOSAL STAGE

Like most projects for virtually any sort of medium, educational projects begin with a proposal document that presents a concept worthy of further development. Good concepts, as mentioned in Chapter 2 and elsewhere, involve not just an understanding of content but also of how that content melds with interactive structure and interface. Although these same concerns are present during the design of educational programs, the proposal stage tends to focus first on other concerns unique to the development of instructional media, especially the need to incorporate into the proposal a clear statement of instructional goals. In this regard, instructional designers conduct a particular type of research known as needs analysis.

Needs Analysis   One of the first tasks of the writer/designer of an educational program is to determine the needs of the people for whom the program is intended. Toward this end, a *needs analysis* is conducted. Although the research in a needs analysis can be quite formal and extensive, involving statistics and survey forms, it can also be accomplished through informal interviews and observation. In either circumstance, the questions that designers ask follow the old journalistic paradigm regarding the six fundamental questions to ask of any interview subject: Who? What? Where? When? Why? How?

The first two questions—Who is the audience, and what do they need to know—are probably the most important, or at least they create the context for the others. These are the starting points, and from them the designer builds an audience profile and composes a content outline.

In making an audience profile, the writer/designer must consider not only the students who will be doing the learning, but also the teachers and administrators who will be using and administering the program. Consider, for example, a language fundamentals program designed for use by fourth graders in inner-city classrooms in California. In order to create a truly useful program, the designer must understand the student population, not only in terms of their educational background and needs but with respect to certain cultural imperatives as well. The designer obviously must research the students' use of language, not just in the classroom but also in the schoolyard and at home. The designer will want to know what types of things the students do for entertainment, what kinds of hobbies they have, what kinds of music they listen to, what kinds of neighborhoods they live in, and what things are signs of prestige and meaningful achievement in their world.

The writer/designer must also become familiar with the other half of the audience, the teachers and the administrators, who will also have close contact with the program and ultimately will be held responsible for its content. Though certain areas of commonalty will exist between students and teachers, there will also be areas of conflict. The goals and values of a Chicano child in East Los Angeles, for example, might be very different from those of the middle-aged, middle-class teacher who runs the classroom. In turn, the concerns of teachers might be very different from those of administrators, who must attend school board meetings and meet with parents and politicians.

In addition to understanding *who* is being taught, the writer/designer must also understand *what* is being taught. In other words, in addition to studying the audience, the designer also must study the content. This portion of the needs analysis consists of consultation with content experts and teachers, as well as research into books, videos, and other presentation materials. The goal is to define and limit the subject matter further, to determine which skills (and basic principles underlying those skills) need to be taught, and to develop a course plan. Content development usually begins in the form of establishing general goals, then continues more specifically as the design process continues.

Other questions involve exactly *when* and *where* this program will be used. In other words, what is the nature of the learning environment? Is it noisy and crowded? Will the students work alone in front of the computer, or will they gather in groups in front of workstations? Is the majority of work to take place during school hours, or is there an after-hours component? Related to such matters is the question of implementation, the matter of *how* the content is to be delivered, including technical questions regarding platform and format. In order to know what is possible to do in terms of multimedia, designers need to know what kinds of computers are being used and what their capabilities are. For schools having fixed budgets and outdated equipment, this is particularly important. In other instances the designers might have the luxury of choosing the type of platform that is best suited for the educational task at hand.

The final question concerns motivation. In the traditional terminology of instructional design, motivation addresses the reasons *why* students are interested in learning. One of the first things the designer must consider is whether learners are coming to the material voluntarily, or whether the program represents a new way of doing things to which learners might have some resistance.

In either event, the designer needs to determine what learners hope to get out of the experience. Why are they engaged in this situation? What are their conscious goals, and their unconscious goals as well? To address any of these questions, the designer needs to understand the audience thoroughly.

The designer also needs to know the expressed and hidden motivations of the project's teachers and administrators. In the situation previously discussed, the expressed motive of everyone involved might be to help inner-city students improve their English-language skills while familiarizing them with computer technology that might otherwise be unavailable. The unexpressed motivation might result from the fact that the school's administrator is under pressure to raise test scores, or perhaps the school was awarded a technology grant. Teachers may see the implementation of such a program as a way of keeping unruly students occupied and engaged, whereas students' initial interest may not center on reading at all, but in the program's animated sequences that are reminiscent of television and video games. In order to appeal to both audiences and design a meaningful program, designers need to consider not just idealistic intentions, but practical matters as well.

The Mission Statement   Once the needs analysis has been conducted, instructional designers often use that analysis to compile a document, called a mission statement, that states the goals and objectives of the proposed program. A mission statement is essentially a concept document that is similar in scope to those mentioned in Chapter 6, though its audience is not quite the same. Whereas a concept document in that case was written largely for a game development team, a mission statement is often reviewed by the educators themselves. It is a way for the development team to stay in touch with the audience, to ensure that they are creating a program that suits the users' educational needs.

The exact content and structure of a mission statement varies according to the project and the audience, but often it is an early-stage document intended primarily to clarify the program's general objectives and specific instructional goals. The writer/designer studies the needs analysis to determine these objectives and then presents them in the mission statement. For the inner-city reading program we have discussed so far, a list of general objectives might look as follows:

PROGRAM OBJECTIVES

- To provide for students an immersive educational experience that reinforces their reading abilities while teaching elements of basic grammar
- To teach underlying foundational skills regarding sentence and paragraph structure
- To increase students' technical literacy skills and to stimulate their interest in using computers
- To appeal to students of a variety of cultural backgrounds
- To provide teachers with a computer-based reading program that can work in concert with the current curriculum
- To create a program with which teachers are comfortable and that requires minimal technical know-how to use and support
- To provide a way to quantitatively measure students' test scores

Notice that the first part of this list focuses on the students, whereas the rest focuses on issues of concern to faculty and staff. This division reflects the dual audience discovered during the needs analysis.

In addition to the general objectives, the mission statement must address specific *instructional goals*. The instructional goals (or instructional objectives, as they are sometimes called) specify the exact skills the students need to learn and in many ways are the meat and potatoes of the mission statement. Although general objectives are useful in establishing tone and context, specific instructional goals are absolutely essential in identifying the skills to be taught. Without a precise statement of instructional goals, the project's focus is likely to drift, particularly when it comes to matters of assessment. It is very difficult to assess the success of an educational program if it is not clear what skills are being taught.

For the inner-city reading program, a partial list of specific instructional goals might be arranged as follows:

INSTRUCTIONAL GOALS

Primary Goal: At the end of this program, students should possess the skills needed to pass the reading/comprehension and grammar tests in the Statewide Comprehensive Examination to be administered at the end of the school year. More specifically:

1. Students should be able to name and identify all eight parts of speech: noun, verb, adjective, adverb, preposition, article, conjunction, and interjection.

2. Students should be able to identify the various parts of speech as to their syntactic function within a sentence.

3. Students should be able to write both simple and compound sentences.

4. Students should know the fundamental structure of the paragraph and understand the function of a topic sentence.

5. Students should be able to read a story and then write a paragraph summarizing the story's theme and plot. The paragraph should have a topic sentence.

The primary goal, explained at the top of this list, functions as the overriding goal to which everything else must relate. It is the reason for the program, and in many ways it answers questions regarding need in that students must learn the skills taught in this program if they are to advance to the next grade. The enumerated list following the primary goal consists of specific instructional goals that must be attained in order for the primary goal to be achieved.

The listing of the instructional goals is in many ways the primary purpose of the mission statement, but it is not the only purpose. The following outline includes many of the major topics that may be covered in a mission statement.

1. Overview
   General description of the program, audience needs, and subject matter; statement of general purpose

2. Goals and Objectives
   Descriptions of goals and objectives, with particular focus on instructional goals

3. Content Elements
   Major content areas to be covered; key skills and concepts to be taught

4. Course Design
   Educational approach, course outline, and media elements

5. Technology Issues
   Delivery platforms and formats; preliminary comments on the interface

6. Other Items
   Work schedules, materials to be consulted, appendices, and so on

Mission statements, like most forms of written communication, vary in structure and level of specificity. In general, their focus is on goals and objectives rather than on program specifics. The specifics come at the next stage in the process, as the designers interact more closely with the content and come to a better understanding of the design shell and the appropriate instructional strategies.

CONTENT AND STRUCTURAL DEVELOPMENT

After the instructional goals have been determined, the next stage involves detailed development of content and interactive structure. These tasks begin with a serious examination of the content itself. The results of this work are ultimately compiled into a treatment document that integrates content and design and describes how particular educational objectives will be accomplished on-screen through particular interactive activities.

Content Development   Even though content development starts with the needs analysis, it often does not begin in earnest until after the mission statement has been drafted. The ultimate goal is to come up with a course outline that will work in an interactive format. As previously noted, this task involves research into the current teaching situation, including consultation with experts in the field.

Good teachers have a certain gift, an intuitive feel for making a subject matter accessible and explaining difficult concepts. Good instructional designers may also possess a similar kind of intuition. Even so, there are certain systematic ways of approaching content development.

The content for a given course consists of the body of knowledge that instructors seek to teach. This body of knowledge consists of three components: (1) the primary *skills*, or conceptual knowledge, that teachers strive to convey (otherwise known as the instructional goals); (2) the material, the case examples and basic raw data; and (3) the various methods teachers use to analyze and present the raw material to students. As a result, the instructional designer's task during content analysis is to examine the overall body of knowledge and organize it according to the skills, materials, and teaching methods involved.

In the inner-city reading program, the primary skills being taught are reading and comprehension. The primary instructional goal might be to have students read a story and then write a paragraph or two describing its main point. Associated goals are the ability to write simple and compound sentences and arrange them logically within a paragraph. The raw materials used in such a course might include storybooks and other assigned reading material, plus some sort of workbook on grammar and sentence structure.

But what about methods? How do teachers use raw materials to convey the desired skills? There are two fundamental approaches. One is through immersion: by fully engaging students in the relevant activity. In contemporary jargon, this approach is often referred to as hands-on learning. The other method is through analysis: by breaking a subject matter into its component parts and then examining those parts. This approach involves developing various classification schemes, including process analysis.

Within the teaching profession today, educators debate whether immersion or analysis works more effectively. The truth is that the two approaches are intimately related and often used in tandem. This is true even in elementary reading classes. Teachers commonly use immersion techniques, surrounding the students with books, having them silently read along while the teacher reads aloud, helping them to sound out the words, discussing the story afterwards. In addition, teachers also use analytical techniques and classification schemes, including teaching phonics, grammar, and syntax as a means for internalizing the various structures of human discourse. In the teaching of writing, teachers rely on process analysis, taking students through the brainstorming, outlining, rough draft, and editing stages. Such combined use of immersion and analytical techniques is extremely commonplace and seems to complement the cognitive process naturally.

After the designer understands the basic skills sets and teaching methodologies, the next step in content analysis is developing a course outline. This task involves determining a sequence for the teaching of particular skills and concepts.

In a grammar course, for example, the parts of speech are usually taught before grammatical function; students then proceed to learn about sentence structure, then paragraphs. Most courses are arranged in a similar way, so that the attainment of certain foundational skills leads to the development of other, more complex skills. In textbooks, related skills and concepts are grouped together in chapters. In a classroom situation, in which a number of materials might be used in teaching a given concept, teachers refer to such groupings as modules.

A *module* is a series of activities based around a particular skill set. A module on the noun, for example, might take one week to complete. Its materials might include a storybook, flash cards, charts, a video, and perhaps even games that students play in class. The teacher would present the material, and then the students would engage in a number of exercises, activities, and homework assignments designed to internalize the concept. All these activities would be followed by a review and perhaps a test. After the class had finished the module on nouns, it would then go on to another module, perhaps one on verbs.

Modules are often used in multimedia education as well. In developing a course outline, one of the designer's tasks is to break the material into modules and then develop multimedia activities for all of them. Each module will have its own instructional goal. In this case a module goal might be to help students identify the nouns in an English-language sentence.

In general, modules follow a predictable structure: introduction of material, presentation of concepts, practice activities, review of concepts, and assessment. The function of the introduction is essentially to warm-up the audience—to introduce the material in a way that is relevant and interesting to students. The introduction also acts as a transition to the main presentation, in which the basic educational concepts are explained. The practice activities give students a chance to apply what they learned and get feedback on the results. The review provides students reinforcement of what they have learned. Assessment, or testing, is a way of evaluating how well students grasp the necessary concepts and skills. Figure 7.1 contains a sample outline based on the structure described here for a module that teaches kindergartners about shapes.

Assessment usually takes place at the end of every module, and then at the end of the entire program as well. Even though formal tests are not a part of every learning situation, they are usually present in some form or another. Often they are quite crucial, especially in circumstances in which skills are cumulative and it is necessary to master one skill set before proceeding.

When the entire content analysis is finished, the findings are frequently compiled into a content document that summarizes all of the elements we have discussed: basic skills, materials, teaching methods, course outline, and assessment. The purpose of the content document is to define the course's subject matter and identify the primary concepts and skills that students must learn. The content document lists the teaching materials and teaching methods involved. It offers a course plan suggesting the substance of individual modules, and it outlines an approach for testing.

Not all multimedia educational programs incorporate formal classroom methodology. Most often, multimedia is used as ancillary material in the classroom; it supplements more traditional educational activities. In such cases, the

Figure 7.1
Outline of an
Interactive Module
to Teach Youngsters
About Shapes.

SHAPES
Instructional Goal: To teach kindergarten students to name, manipulate, and recognize the fundamental characteristics of basic geometric shapes: the square, the rectangle, the circle, and the triangle.

INTRODUCTION OF MATERIAL
Purpose: To catch students' attention and introduce the subject matter. Opening screen sequence will feature an animated cartoon with several characters: Mr. Square, Ms. Circle, and so on. When students click on characters, each shape will introduce itself.

PRESENTATION OF CONCEPTS
Purpose: To explain the various shapes in detail. Toward this end, subscreens for each shape will teach students that shape's fundamental characteristics. The screen about squares, for example, will teach that the square has four sides of equal length.

PRACTICE ACTIVITIES
Purpose: To give students practice manipulating shapes, learning their characteristics in the process. They can use a paint screen and simple animation program to draw, color, and animate various shapes. Another screen will allow students to combine shapes, using squares and circles to make people, triangles to make trees, and so on.

REVIEW OF CONCEPTS
Purpose: To reinforce the concepts learned. On the review screen, the characters from the opening animation reappear. Students examine common household objects and learn how their forms reflect the basic geometric shapes (a book is a rectangle, a penny is a circle, and so on).

ASSESSMENT
Purpose: Assessment and testing. Students are asked to sort shapes into categories. This activity might take the form of a game in which students sort the shapes as they go by on a conveyor belt. The game will be scored, and students will be able to play it as often as they wish.

exploration of subject matter may be open-ended, testing may be omitted, and the modules may not be arranged sequentially. Even so, many of the principles we have discussed still apply, particularly when it comes to developing content around an educational theme.

Interactive Structure   After the content has been determined, the designer must devise an interactive structure that supports the course plan. As in other forms of multimedia, designers conceptualize the design shell of an educational program through the use of design metaphors and classification schemes. Even though we have already described this process in detail in Chapters 2 through 5, it is worth repeating here that the creation of the design shell involves integrating the three primary design elements: content, interactive structure, and interface. In a good educational program, these elements are integrated in a way that serves not the design itself, but the instructional needs of users.

In addition to designing an overriding structure, the designer of educational software must also generate individual interactive exercises appropriate to the subject matter. Thus instructional design for interactive media involves not only conceptualizing the design shell but developing individual instructional activities as well. In both respects, the most common technique for generating interactive structures is by using familiar, traditional learning activities as design metaphors.

For example, one way children learn to read involves storybooks. Typically, an adult reads a story to the child, occasionally pausing to answer questions about the pictures or the words, to embellish the story, to add background and additional descriptions, or to relate the story to something else in the child's experience. As we noted earlier, this storybook experience has been used frequently as the primary organizing metaphor in adapting children's books to the computer screen. In this context, the computer assumes the role of the adult in providing the storyteller's voice. Instead of asking the human reader questions about the pictures and words, the child uses the mouse, pointing at the screen to hear words repeated or to learn other information. Broderbund, Scholastic, Computer Curriculum Company, and numerous other software companies have used this methodology as the basis of the design shell for multimedia programs that teach beginning reading skills.

Similarly, classroom activities can function as models for individual activities that occur within the context of the larger design shell. Spelling bees, flash-card exercises, quiz shows, and playground games can and have been adapted to the computer screen. Activities that involve matching and sorting are particularly adaptable to the computer. Teachers frequently use sorting bins to teach classifications or manipulate counting sticks to teach math. Such activities can be mimicked by the point, click, and drag functions of the mouse, making it possible to move objects from one part of the screen to another. Figure 7.2 depicts an activity in which students use the mouse in this fashion.

A second way of designing interactive structures is by focusing on the processes and tools used during a particular activity (see Figure 7.3). The emphasis here is on the act of construction itself—on getting the students to use the computer to create and assemble. In this type of program, the focus might be on getting students to learn about reading by creating their own storybooks. They might be given a landscape on a computer screen, a choice of characters to place on that screen, and finally an opportunity to write a sentence describing the scene they have created. There are in fact several programs of this nature, as well as programs that offer students the chance to compose music, make photographic essays, or solve equations. One interesting variation on this approach involves the use of a media lab in which students play the part of reporters who view video clips on a given subject, then create a multimedia report to demonstrate what they have learned.

A third strategy for developing interactive structures involves developing immersive environments that mimic real-world places and situations. Such envi-

Figure 7.2
One Use of a
Familiar Activity as
a Design Model. In
this activity, students
learn about fractions
by playing the part
of a cafeteria
counterperson,
scooping up portions
of food to serve to
waiting students.
(*Jumpstart First
Grade,* Knowledge
Adventure.)

ronments are particularly useful in those types of training in which it is too expensive or dangerous to place novice trainees in real-life situations. Pilots, for example, sometimes do their initial training on multimedia flight simulators, and the same is becoming increasingly true for operators of certain types of machinery and for extremely specialized professionals. Programs developed for surgeons enable them to practice their skills on imaginary patients, and programs for architects allow them to walk through their creations before they are even built.

In addition, the world of computer games can also provide an inspiration in the design of educational programs, either as overriding design metaphors or as models for individual activities. Arcade, strategy, and role-playing games have all been used successfully for educational purposes. The popular Muncher games, used to teach students about numbers and spelling, are in fact arcade games based on the legendary *PacMan.* In the adventure-style game called *Oregon Trail,* students learn about nineteenth-century U.S. history while playing the role of a wagon master trying to guide a wagon train across the American West (see Figure 7.4). Success in such games involves the strategic manipulation of inventory, just as it does in more entertainment-oriented games. The difference rests in the educational intention, which is reflected both in the accuracy and depth of historical detail and in the type of tasks the students must complete in order to advance in the game.

No matter the type of development technique used, the primary concern of the designer at this stage is integrating content and structure. This task involves

Figure 7.3
The Use of
Processes and
Tools as Design
Models. In this
activity, students
learn about color
and shape using
tools and processes
that mimic those
found in the real
world. This activity
also teaches
students how to use
paint programs on
the computer.
(*Jumpstart First
Grade,* Knowledge
Adventure.)

establishing an overriding design metaphor or classification scheme and individual activities that work within the context of the larger organizational pattern. Sometimes the approach is to divide the content into a series of sequential modules, each with its own objectives and activities. Other times the modular approach will be abandoned for a full-scale immersive activity that is more narrative or game-like in structure. Regardless of the approach used, one of the designer's primary concerns is to integrate the computer's interactive capabilities into the program; otherwise, the instructional task might be best accomplished in other media.

Writing the Treatment　　The design process in instructional media is not a solitary enterprise. It involves the collaboration of people from various disciplines. This is particularly true when it comes to developing interactive programs that must not only embody instructional goals within the context of a course plan but must also successfully use graphics, sound, and text on the computer screen. Thus the interaction among educators, media people, graphic designers, and writers is likely to be most intense at this portion of the design process.

　　The writer at this point can have several tasks: to be a source of creative ideas for actively shaping the design process and integrating form with content; to help facilitate communication across disciplines; and to act as a creative synthesizer, bringing together the course objectives, design strategies, and interactive exer-

Figure 7.4.
The Use of an
Immersive
Environment to
Structure
Interactivity.
Oregon Trail II
teaches students
about the American
frontier by placing
them in charge of a
wagon train on its
way to the Pacific
Northwest. This
program uses
techniques from
adventure games
and also requires
students to engage
in dialogue with
various characters
along the way.
(*Oregon Trail II,*
MECC.)

cises into a document that serves as a touchstone for other members of the design team. The purpose of this document is to provide an overview of the program that summarizes each section's main educational objectives and outlines the major activities. Regardless of whether this document is called a treatment, a program outline, or a design document, its function is the same: to provide a summary of the major educational and design components of the program.

Consider the following excerpt from a treatment for an educational reading program:

SAMPLE TREATMENT

MODULE ONE: NOUNS

MATERIAL

The core of this lesson is the children's storybook *My House.* The focus of the book is on objects commonly found around the house. The writing style is simple, involving short declarative sentences that emphasize the common household objects themselves.

The pages of *My House*—including text and pictures—will be digitized and presented in a multimedia environment. Each page will be enhanced with graphic and audio material, including animations. These enhancements will include educational activities directed at teaching students the concept of nouns.

PRIMARY OBJECTIVE

To teach students to identify nouns in an English-language sentence

ACTIVITIES

I. **Background Activities**

    Before entering the book itself, students must first enter certain background activities that will be presented as button choices on the opening screen.

    1. Video—A video in which children are seen naming household objects and identifying them as nouns.
    2. Definition—An interactive exercise reinforcing the classic definition of a noun as a person, place, or thing.
    3. Match Game—A drag-and-click activity in which students use rebus pictures to match nouns with the objects they represent.

II. **Storybook Reading of *My House***

    In this phase, students work their way through the story page by page. Each page begins with an audio reading of the text. After listening to the text, the students explore the page, clicking on icons and hot spots. Activities include:

    1. Mandatory reading of the text by instructional voice.
    2. Obligatory cut regarding the "featured noun" on each page.
    3. Dictionary Icon—provides audio/visual definitions at the student's request.
    4. Hide and Seek Button—allows students to look behind certain objects for "hidden nouns," which they can then "paste" into their journal and compose into sentences.
    5. Secret Passages—hot spots that take students to other parts of the house.
    6. Label Hot Spots—audio/text hot spots that provide on-screen text labels synchronized with an audio pronunciation guide.

III. **Review**

    After students have finished the book, the following activities are presented on the test-and-review screen.

    1. Eat the Noun—a Muncher-style game in which students seek out nouns in an underground maze.
    2. Quiz Show—a TV-style trivia game about nouns, hosted by a character from the book.
    3. Make Your Own Book—an activity in which students use rebus pictures to make their own multimedia show about nouns.

The treatment excerpt presented here is in outline format; as such it functions as a preliminary script from which a more detailed script will be written later. It will also serve as the basis for discussion among various members of the design team. Consider, for example, the Definition exercise listed as Item 2 under Background Activities. As described in the treatment, this activity is still in a very early stage of development. All the designers know so far is that they want an activity that will communicate the definition of a noun; the exact nature of that activity still needs to be determined. Is this a point-and-click activity? Does it involve pointing at and dragging objects and then placing them next to the text, or should it involve collecting objects into a container of some sort?

Suppose that the writer/designer decides to take an approach that is reminiscent of television game shows in which a player chooses squares one at a time from a large grid. When a square is chosen, it flips over to reveal a picture or word underneath. Players score points when they find pictures and words that match. Even such a relatively simple design requires consultation within the design team. Numerous specifics need to be worked out among the graphic designer, the interface designer, the programmer, and educational experts. This simple activity must be constructed not only so its content is appropriate to its place within the lesson plan, but so it also matches the tone of the program and appeals to the audience at hand. This process of bringing the treatment to the next level of specificity is called script design.

SCRIPT DESIGN: CREATING A PRODUCTION BLUEPRINT

Once the treatment has been written, the next stage is to create a production script that acts as a template, or blueprint, for the program. The production script not only mentions educational objectives; it also specifies and describes text elements, audio, graphics, and any animations or video. In addition, it must be written in such a way as to make the interactive possibilities clear to the programmer. For this reason, writing the production script often involves consultation across the design team.

The following script, for example, delineates the different actions which the user can take on Screen #2A in a program about nouns. Before taking a look at this script, however, it might be best to discuss some of the formatting conventions it uses.

The script begins by stating the screen's educational objective, describing the activity, and listing the media elements. Then the script delineates the computer's response to various user actions; a line of asterisks is used to separate one action from the next.

To clarify the sequence of media events triggered by a user's action, the script takes advantage of certain conventions. General instructions to the programmer appear within double parentheses. AUDIO indicates spoken narration. AUDIO/TEXT describes on-screen text accompanied by a simultaneous audio reading of the text; during the reading of audio/text sequences, the text is visually highlighted word by word. Similarly, the phrase "illumination sequence" refers to the sequential highlighting of graphic images on screen. Illumination sequences are often used to introduce hot spot choices and can be accompanied by audio narration and musical effects.

When reading this script, it is helpful to keep in mind that an interactive script is very different from other scripts in that its purpose is not necessarily to capture the narrative flow of events, but rather to serve as a production template. Reading such scripts—like writing them—takes time and practice.

SAMPLE PRODUCTION SCRIPT

MODULE ONE, BACKGROUND ACTIVITY #2: DEFINITION OF A NOUN

SCREEN #2A

OBJECTIVE
To teach students the standard definition of a noun

ACTIVITY DESCRIPTION
On screen are three pictures, each representing a major class of noun. Students click on these pictures to see and hear how the object pictured fits into the definition.

GRAPHIC ELEMENTS
The screen should be drawn in such a way as to accommodate the following audio/text and graphic hot spots based on images from *My House.*
1. A young girl—Chinese-American, smiling, about seven years old
2. A city—a city skyline, with buildings, bridges, houses, streets, and so on, drawn to look like a friendly place to live
3. A kite—flying in the sky, painted with the image of a dragon

AUDIO ELEMENTS
1. Voice-over narrative
2. Sound effects for illumination sequence

INTRODUCTORY NARRATION
((When the student enters Screen #2A, run the following sequence:))

AUDIO/TEXT: A noun is a word that names a person, place, or thing. ((Pause))

AUDIO: Click on each picture to learn more.
((Illuminate each picture in sequence. Run sound effect in sync with each illumination.))

HOT SPOT #1: THE GIRL
((If the student clicks on the girl, run the following sequence:))

AUDIO: Girl. A girl is a person.
((Pause. Replace the image of the girl with the word "girl," written in large, block letters. Resume audio.))

AUDIO: The word "girl" is a noun.

HOT SPOT #2: THE CITY
((If the student clicks on the house, run the following sequence:))

AUDIO: City. A city is a place.
((Pause. Replace the image of the city with the word "city," written in large, block letters. Resume audio.))

AUDIO: The word "city" is a noun.

HOT SPOT #3: THE KITE

((If the student clicks on the kite, run the following sequence:))

AUDIO: Kite. A kite is a thing.
((Pause. Replace the image of the kite with the word "kite," written in large, block letters. Resume Audio.))

AUDIO: The word "kite" is a noun.

PROGRAMMER: ((If the student fails to click on a hot spot within 30 seconds, rerun Introductory Narration for Screen #2A. The student must press all three hot spots before exiting. Upon exiting, play closing narration and closing music.))

CLOSING NARRATION:

AUDIO/TEXT: All right! A noun is a person, place, or thing.

RUN CLOSING MUSIC (Flourish, upbeat)

A script is reviewed many times before it goes into production—by graphic artists, project directors, content experts, and others associated with the project. Part of the review process often involves the creation of a prototype, or demonstration version of the program, for testing by the audience. In multimedia, as in any presentation medium, a gap can exist between the intent of the program and the audience's experience. The goal of audience testing is to narrow that gap as much as possible.

IMPLEMENTATION AND REVISION

In our earlier discussions of the design process, we noted that the writer/designer's role usually ends once production begins. This is less often the case in instructional media, in which educational concerns often require the continued involvement of an instructional designer both during and after program implementation.

In instructional design, the term *implementation* refers to that period of time during which a fully functional version of an educational program is put into actual classroom use. This may sound like a simple matter, but in an educational or corporate setting, implementation is a major concern. The people who will be using the educational software often need to be taught how to use it. In addition, for training programs involving sophisticated software and hardware, the system will require maintenance and support. All these tasks require active involvement of instructional designers, not only to explain to users how the program functions but also to observe its strengths and weaknesses as an educational tool.

Even after the program is implemented, testing continues. Designers need to evaluate the effectiveness of their programs, to be assured they are reaching the audience with the intended message. Such continual evaluation serves as a guide both in updating existing programs and designing new programs for the future.

There are several ways to evaluate the effectiveness of instructional material, but all involve getting feedback from users. One common technique involves choosing a small focus group of students that is representative of the program's audience as a whole and then conducting in-depth, follow-up interviews at various times. Another technique involves using larger groups of students, survey forms, and statistical analysis. A third approach includes building an immediate response feature into the instructional program so that students can comment on its instructional effectiveness immediately after experiencing it. A fourth way consists of field trials, in which student performance is evaluated both before and after training. Each of these techniques has its limitations. How each is used, and in what combination, depends on the type of training the program involves.

Once the analysis of implementation has been done, it is then time to revise the program. When undertaking revision, the designer must decide whether to revise the content, the instructional approach, or both. When revising content, the designer makes changes and refinements in the instructional material with the goal of making that material more accurate and more effective as a learning device. In contrast, when revising the instructional approach, the designer changes the underlying teaching method. An example of such a change would be to switch from a teaching model based on explanation of process to a model based on immersive training.

Revision is an integral part of instructional design. Good teachers are always refining their efforts, incorporating new materials and presentation techniques. The same is true of good instructional designers. Just as textbooks are updated and revised, so are educational multimedia programs. Revision is not an indication of failure, but part of an ongoing process of improvement.

Models for On-line Education

Even though much has been made of the possibilities that the Internet and the World Wide Web hold as educational tools, they remain relatively unexplored media. Until very recently, the Internet has served largely as a research tool for scientists and other highly specialized faculty; it has functioned more as a repository of information than as an active teaching device.[2] With the increasing graphics capabilities of recent years, however, many universities and other institutions are making plans to move curricula on-line.

Designing educational content for the Internet involves some special considerations relating to both the nature of the interactivity available and media capabilities. Even so, the underlying design considerations are much the same as those in other interactive media. For an on-line curriculum, as with any educational project, the writer/designer begins by developing a list of instructional objectives that should be based on a firm understanding of the audience, its needs, and a content analysis. The next step is development of a course plan. In this regard, the design of individual modules proceeds according to the process described earlier in this chapter: introduction of material, presentation of concepts, practice activities, review of concepts, and assessment.

The on-line environment does require some special considerations, however. One consideration concerns the media capabilities of the Internet; another involves the special sorts of interactive activities available on-line: live chat rooms, conference forums, links to other sites, and the possibilities of real-time interaction between a user and the remote server. A third consideration relates to the history of the Internet as an academic research tool.

All these factors affect the design process. The World Wide Web, for example, first evolved as a text-only medium. Colorful graphic interfaces, a recent innovation, still download only slowly, impeding the flow of information and exhausting the patience of users. An educational designer must work around this fact and must also take into consideration limitations on the use of sound and video. Although until very, very recently sound and video were available on the Web in file format only (which required users to download and watch off-line), now instantaneous playback capabilities exist. However, these capabilities vary from user to user according to their hardware and software set-ups, so designers need to understand that often they are working in a text and picture environment in which video, animation, and sound can be more impediment than advantage.

Similar considerations come to mind when considering the differences in interactive capabilities. In one sense the possibilities are more limited; on the Web, for example, few possibilities for conditional linking exist. In other words, it is difficult to construct links based on a look back at every place a user has already been, with the computer tracking various pathways simultaneously and changing navigational options accordingly. Links instead tend to be hardwired so that navigational options are dictated by the user's immediate location in the interactive pathway. (For more on conditional linking, see Chapter 9.) Despite these kinds of navigational limitations, though, Web links do open up other opportunities. Designers can link their own screens to those created by others, thus expanding the content and scope of their program. The Internet also allows for real-time discussion, display, and updating of information, thus making collaborative projects a possibility. In places where special equipment—such as *ISDN* lines, *T1* lines, and *T2* lines, or fiber-optic wires—is available, video conferencing and real-time voice conversation are also possible.

In its early days, scientists and researchers used the Internet as a place to store and retrieve specialized information and to keep in touch regarding research projects. As the use of the Internet has grown, it has become a vast compendium of information, though a rather disorganized one that lacks a central filing system. Its primary function in education so far has been as a content source. Students and teachers typically go on the Internet in search of content and supplemental material. Very little of this material, if any, was consciously designed for presentation as part of any organized curriculum.

More recently, some institutions have begun developing on-line material with more specific instructional goals aimed at selected classroom audiences. CNN has placed on the Web a package of video clips with discussion questions that high school teachers can download and use in their classrooms. The University of Virginia has a Web site in which anatomy lab students can participate in a simulated frog dissection (see Figure 7.5). A number of virtual museums and other exhibits can be found on-line; although some of them incorporate exercises for teachers to download and use in the classroom, most are more suitable for browsing than as a structured educational experience.

A true model for on-line education will feature classes and learning models that incorporate clear instructional goals from the outset. In this model, interactive programs not only present the material but also provide ways for students to practice skills and use their knowledge. Such a model would also provide opportunities for review and assessment.

In all likelihood, on-line education will not evolve into a stand-alone medium. Rather, it will be used in concert with traditional elements such as books, videos, and classroom contact with students and teachers. From a practical perspective, there are questions about testing and assessment procedures on-line, particularly for formal accreditation purposes. It is difficult to verify, for example, who might be filling in the answers on a multiple-choice test taken over the Internet. In order to implement on-line education, such issues need to be addressed.

More important, though, is the human factor. People learn not just by being exposed to knowledge, but by interacting with other people. On-line discussions, lack the visceral impact created by contact with real-life, in-the-flesh humans talking about ideas. Even so, long-distance education is likely to continue to evolve, supplementing and in some cases supplanting the traditional classroom, its uses being dictated by such factors as convenience, cost, and the willingness of the audience to participate. In order for such programs to be effective, though, writers, designers, and teachers will need to ground the presentation techniques in solid instructional models.

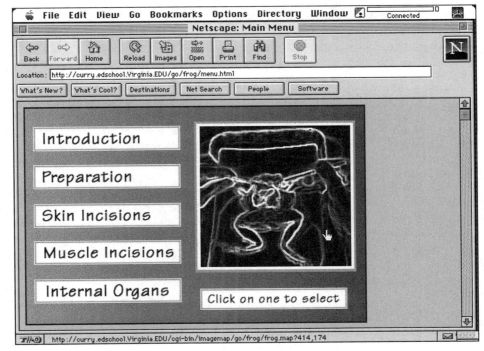

Figure 7.5. Use of a Web Site as an Instructional Tool. This Web site based on an instructional design model teaches students about basic anatomy during on-line sessions focusing on frog dissection. (*Frog Dissection Activity,* Mabel Kinzie, University of Virginia, Department of Instructional Technology. http://curry.edschool. Virginia.EDU/go/frog/ menu.html.)

A Final Look at Educational Multimedia

Even though the process of creating a good interactive educational project can be very complex, it is nonetheless guided by the same simple underlying premise we have mentioned before: Keep the user first. In this context that means identifying the educational needs of the audience and keeping them paramount.

Writers can play many roles in the design process, from researcher to content developer to instructional designer and scriptwriter. Though at times writers may have quite a bit of creative input, the collaborative nature of the medium frequently places them in the role of synthesizer who must focus the script while balancing the diverse and often contradictory requirements of content specialists, media personnel, and programmers. Often writers must play the role of go-between, for they may be the only people (except the producer) who are in contact with all the different personnel and who understand their various concerns. This can be an aggravating position, but it can also be very rewarding, particularly when the cross-disciplinary elements merge in ways that offer fresh glimpses into content and new ways of teaching. Educators are excited about the possibilities inherent in multimedia because it provides a context for exploring old ideas about teaching through experience, immersing students in activities, and providing new ways of learning by doing. As more and more designers borrow techniques from computer games, the possibilities for incorporating gaming and narrative techniques in an educational context grow.

There is, of course, a flip side to the use of computers in the classroom. Part of the allure for administrators is the promise that computers will save time and money while increasing the quality of education. The fear, of course, is that more machines mean not better education, but simply fewer teachers per student. Whether we really should replace teachers with machines, what role computers should play in the classroom, at what point computer training becomes mind-numbing rather than mind-enhancing—these are the kinds of questions that may not get asked often enough. They will certainly never get asked by the computer companies seeking to sell their wares to school administrators.

Whether computers truly save work is another moot point. Anyone who has worked with computers knows that whereas they may eliminate some kinds of work, they create others. Upgrades are always available, downtime must always be endured, compatibility and access problems always frustrate. In the end, the fear is that more attention will be spent tending to the machine than to teaching. Students may learn quite a bit about how the technology works but very little about the subject matter the machines were supposed to present.

If people truly learn by doing, they need also be immersed in the "real" event, not the computerized version. Even though computers are good at sorting information into categories, they are not so good at conveying subtle relationships among ideas. It may be wise to keep in mind that there are simpler and cheaper ways to convey information, including employing a knowledgeable human being to stand in front of a classroom and interact with students.

Even with these cautionary words in mind, however, it is difficult to ignore the myriad possibilities that seem to lie ahead: the ability of the Internet to bring scholars and thinkers together to work on pressing research projects; the availability of sophisticated information resources to people regardless of class or economic status; the excitement that goes hand-in-hand with multimedia development because, by its very nature, it brings together people from diverse disciplines. These alluring possibilities are enhanced by the contemporary notion that new media provide new ways of thinking and perceiving, that human consciousness is moving into a new arena. Whether such notions are wishful thinking or instead the stuff of illusion—and whether or not the educational promises of the new technology will be fulfilled—are questions that only time can answer.

SUMMARY

The instructional design process often begins with a study of the audience. This study, referred to as a needs analysis, takes into consideration such audience factors as motivation and demographics (including age, social background, and other social factors). The needs analysis also establishes the instructional goals concerning the particular skills and concepts the program aims to teach. A concise statement of the instructional goals, together with an analysis of the audience, are then compiled in a concept document known as a mission statement.

After the instructional goals have been clarified, instructional designers focus on the content, arranging the course materials and developing a course plan. The purpose of the course plan is to develop a sequential outline for the presentation of the educational material. Course plans can be broken down into individual modules, each of which focuses on a particular skill relevant to the overall goal. As such, each module has its own instructional goal. Modules are typically structured in a sequential way, involving the introduction of material, presentation of concepts, practice activities, review, and assessment.

Instructional design for interactive media also involves the determination of a design shell. As in other forms of multimedia, designers conceptualize the design shell of an educational program through the use of design metaphors and classification schemes. After the design shell has been determined and activities generated, the writer/designer produces a written treatment that forms the basis for the production script.

As designers adapt this process to different types of interactive media, they need to keep in mind the media capabilities of the format in which they are working. The movement toward on-line curricula is challenging in this regard, particularly as the Internet evolves from a supplemental content source into a presentation device. Fully realized curricula for interactive media, like curricula for any media, depend on the ability of the writer/designer to develop workable models that embody clearly defined instructional goals and objectives.

EXERCISES

1. Choose a subject matter for an educational multimedia program. (A few possibilities: HIV awareness; the cultivation of roses; modern cinema; the history of religion.) Do a thorough needs analysis and present your analysis to the class.

2. Based on the needs analysis conducted in Exercise 1, write a mission statement containing general objectives and specific instructional goals.

3. Write a content document that includes a course outline and module summaries.

4. Develop some specific interactive exercises for your program.

5. Write a production script for one of the interactive exercises developed in Exercise 4.

E N D N O T E S

1. Walter Dick and Lou Carey, *The Systematic Design of Instruction* (Glenview, IL: Scott, Foresman, 1985); R. Kaufman and F. W. English, *Needs Assessment: Concept and Application* (Englewood Cliffs, NJ: Educational Technology Publications, 1979); R. M. Gagne, *Conditions of Learning* (New York: Holt, Rinehart & Winston, 1977); D. A. Payne, *The Assessment of Learning: Cognitive and Affective* (Lexington, MA: Heath, 1974); L. J. Briggs, *Handbook for the Design of Instruction* (Pittsburgh: American Institutes of Research, 1970), 93–162; R. M. Gagne, W. Wager, and A. Rojas, "Planning and Authoring Computer Assisted Instruction Lessons," *Educational Technology* 21, no. 9 (1981), 17–26; R. A. Reiser and R. M. Gagne, *Selecting Media for Instruction* (Englewood Cliffs, NJ: Educational Technology Publications, 1983); J. D. Russell, *Modular Instruction* (Minneapolis: Burgess Publishing Co., 1974); and J. P. DeCecco, *The Psychology of Learning and Instruction: Educational Psychology* (Englewood Cliffs, NJ: Prentice-Hall, 1968), 54–82.
2. Ed Kroll, *The Whole Internet: User's Guide & Catalog*, 2nd ed. (Sebastopol, CA: O'Reilly & Associates, 1994), 13–21.

8

Scripting
for Business
Programs

The glitter may be in entertainment, the content in education, but the money is in business. Corporate training, sales, advertising, and promotional venues—these are among the prime concerns of commerce, its daily bread and butter; the same holds true for interactive media. Game designers may pioneer in matters concerning interface, and educational designers may pioneer in the adaptation of content to interactive structure, but it is designers working for businesses who put these techniques to their most practical applications.

It should be no surprise, therefore, that the design process for business programs follows a familiar pattern. As in education—or in games, for that matter—the writer/designer looks first into the needs of the audience, then conducts research into the content. A concept document is prepared, and that concept document eventually evolves into a production script. The primary differences rest in the needs of the audience and in the type of messages that need to be communicated.

We have already mentioned that in educational programs, the designer must consider two audiences: teachers and students. A similar thing is true for business programs. The designer must consider not only the audience a client intends to reach, but also the client's needs.

In other words, a company may have a particular audience it wants to reach with a particular message, and to convey that message effectively the designer/writer needs to know the company's vision of itself. When designing for established businesses, certain perimeters and restrictions may exist. Slogans, logo style and placement, the use of color, and the like are often governed by internal convention. The same may be true for presentation style and content. Such restrictions may seem arbitrary, but nevertheless they are very much part of a company's public identity. Companies, particularly large corporations, are extremely careful about how they present themselves. Additionally, a company's internal culture and politics may affect the style and content of any given program. The designer must be aware of all these vagaries, and more.

In this chapter we will discuss the primary features of three types of business programs: advertising programs, informational programs, and training programs. Then we will take a look at questions of format and the scripting process itself.

Advertising Programs

Advertising programs are designed to sell a product or service. As such, they are undertaken for entirely different reasons than entertainment and educational programs, with entirely different audiences in mind. As a result, they pose special challenges to designers.

Audiences for games and educational programs usually come to the program possessing their own motivations. In the case of games, the audience is looking for an immersive and largely vicarious experience that challenges their intuitive and emotional abilities in a controlled environment. In contrast, the audience of educational programs seeks knowledge and enhancement of particular skills with specialized applications in the real world. Audiences of advertisements may share some of these characteristics, but there is a fundamental difference: The audience of an advertising program has not yet made an emotional or intellectual commitment to the message being delivered. They are withholding judgment; they are browsing, sometimes with serious intent but oftentimes merely with curiosity.

The purpose of an advertising program is not simply to present a product and its attributes; the purpose is to make a sale. Toward that end, a designer has two goals: to make the customer want the product—to stir the customer's desire—and then to get the customer to act on that desire.

One of the more common design metaphors used in advertising programs is the brochure or shopping catalogue. In such a model, products are grouped by category. A customer first chooses a category, then scrolls through products one at a time, clicking on those of interest to examine them more thoroughly. In addition, such programs often include search engines, which allow customers to type in the name of the object of their desire, thereby going directly to it.

A major consideration in designing an interactive advertising brochure is its structure. Although game players and certain educational audiences may tolerate or even enjoy roaming around very ornate interactive structures, this is far less likely to be true for consumers, whose primary interest is to learn about a product and decide whether or not it suits their needs. For this reason, the interactive structures of mainstream advertising programs tend to be simple, relying on classification categories or straightforward design metaphors. According to conventional wisdom, their branching structures should be simple, hierarchical, and largely horizontal, designed so that users can get quickly in and out of the material.

However, this is not always true, particularly when the audience is younger, or has technological savvy, or enjoys playing with interactive media. For products like computer games, certain kinds of movies, or interactive media, the audience in fact expects a certain amount of innovation in the design. In such cases, the interactive structure may be considerably more playful, with more complex branching and perhaps intricate loops.

Another structural consideration concerns the arrangement of material on individual screens. Here, too, interactive designers take their cues from other media, including television and print, in which the primary focus is on image. In advertising, as in life, the first impression is often the most important. It is the first impression that lingers and informs all that follows. Thus, when a user first encounters products on screen, they are often presented dynamically, up front, dominating the screen. That presentation is often accompanied by a key phrase —a line of text or audio—that somehow imbues the product with a certain spirit or trait that separates it from other similar products.

We all know about such tag lines, phrases like "Coke: The Real Thing," "Nike: Just Do It," and "United: Fly the Friendly Skies." Such phrases are designed to create product identity, but they need not always focus on abstractions. They can also focus on particular attributes such as price, speed of service, or other mechanical and physical features that separate a product from its competitors.

Not only must the opening sequence in interactive advertising create a good first impression consistent with the product's image; it must also draw the user deeper into the program (see Figure 8.1). This can be said of the main navigational screen as well. The opening sequence and the main screen serve as introductions to the subject matter.

In advertising, introductions are usually tied to *sales concepts*, phrases or ideas that embody an attribute the advertisers hope will help sell the product. Among the classic examples of sales concepts are the association of sex appeal with cigarettes, the association of elegance with expensive automobiles, and the association of youthful defiance to loose-fitting jeans. In each of these examples, the product is associated with a sales concept designed to appeal to a specific audience. The sales concepts may not be expressed explicitly in the ad, but they nonetheless permeate the ad's visual and verbal elements.[1]

Main screens for interactive media are often designed according to a similar philosophy. The main screen of the Delta Airlines Web site is a good example (see Figure 8.2). Here, navigational options are presented clearly and straightforwardly beneath a background of an imaginary city composed of images from the various places to which Delta flies. The underlying imaginative concept is to associate the airline with exotic and appealing travel destinations.

An interactive presentation may also contain other substantive material, including written articles and in-depth descriptions, but such information is usually not the focus of the main screen. The substantive information, instead, will be down another layer in the interactive structure, placed in hot spots or subscreens. The designer's first mission is to whet the user's appetite; after that has been done, issues of substance can be addressed. This is not unlike the approach used in traditional advertising brochures that flashily present the product and tag line on the cover and place the substance inside. Once the audience's attention

Figure 8.1
Use of Main Screen
to Establish
Company Identity.
This Web site very
clearly announces
both the name of the
company and the
service the company
provides. (*Atomic
Vision,* http://www.
atomicvision.com/
rotary.html.)

Figure 8.2
A Main Screen
Implying Simplicity
of Underlying
Structure. On the
head banner, this Web
site announces "You're
Just One Click Away
from Anywhere in the
Delta Skylinks Site,"
suggesting the use of
a shallow, streamlined
interactive structure
designed for ease of
use. Major classifica-
tion categories appear
as icons in the water
beneath the cityscape,
then again at the
bottom of the page.
(*Delta Airlines Web
Site,* http//www.
delta-air.com/index.
html.)

has been captured, the focus can shift to the content, to converting the consumer's desire into need.

One of the key principles of advertising is to persuade consumers that they *need* the product to stay healthy, to stay competitive, to survive in the world, or the like. In one form or another, this effort at persuasion continues throughout an interactive program. Toward this end, writers and designers use many of the techniques that are effective in entertainment and educational multimedia.

To guide consumers through a presentation, designers sometimes make use of a guide figure, a character who acts as a mentor of sorts, on the one hand helping users to navigate the program but also engaging in gentle (and sometimes not so gentle) persuasion. One example of a guide figure used in this way is the woman who appears on *2Market*, a multimedia shopping catalogue put out in association with America Online (see Figure 8.3). The function of the guide figure is both to diffuse users' anxiety over the technology and to encourage them to buy from the catalogue. This young woman—cheerily middle class, upbeat, and seemingly forever willing to shop—apparently represents an idealized vision of the program's target audience.

Another concern in interactive advertising is the balance between text and image. For the most part, text plays a supporting role. When sharing the screen with graphics, its usual role is to augment the image—to call attention to the abstract qualities the advertisers want to convey or to point out particular sales features. In this situation, text often has a similar role to text in magazine advertisements, and the writing skills needed in these circumstances are the skills of

Figure 8.3
Use of a Guide Figure in a Shopping Catalogue. This young woman tells shoppers how to make their way around this interactive shopping brochure on CD-ROM. (*2Market*, 2Market, Inc.)

an ad writer: brevity and conciseness. Even so, text can play more than a supporting role in interactive media. The balance sought between text and image has much to do with the product, the ad, and the intended audience. Before they make the final purchase, consumers often want to know the details of a product's features. Specialty articles, essays, or news briefs still remain the best ways to provide certain kinds of information, and many advertisers make it a point to include such material in their interactive presentations. Rarely, though, is lengthy written material the focus of the main screen.

The purpose of every facet of an interactive advertising program is to stir users' interest and to create desire. This is as true of each individual screen as it is for the program as a whole. After desire has been stirred and the need for the product established, it is crucial that the designer provide ways for consumers to act on their impulses. After all, a consumer's impulse to buy passes quickly. Here, too, interactive presentations borrow from the techniques of magazine and catalogue advertisers by providing order forms, telephone numbers, and other features to make it convenient for customers to act on their impulses. Interactive presentations also make use of obligatory cuts and transitional screens that give users a chance to order the product via modem, immediately, while the impulse remains strong.

Obligatory cuts are useful ways of reminding users of product features, of presenting sales pitches, or of giving customers a last chance to order before signing off. They provide phone numbers, addresses, coupons to print out, or direct modem connections to the dealer. As mandatory viewing, they give designers a powerful tool that must be used with caution. When trying to win over potential customers, it may be unwise to use an obligatory cut to entrap them in a presentation they have already chosen to leave.

Most advertising, no matter the product, shares a similar purpose: to persuade the consumer to buy. Interactive advertisers, like those in traditional media, use sales concepts based on rhetorical and associative patterns of persuasion. Designers, too, must make decisions about various aspects unique to the medium. In many cases they strive for simple interfaces: broad and shallow branching structures, text messages that are quick and to the point, and opening screens that focus on image as opposed to substance. These tend to be hallmarks of interactive advertising, particularly for mainstream advertisers appealing to an audience whose primary interest is not in the technology but in the product. As the audience becomes more familiar with the conventions of interactive design, however, designers tend to take more liberties. They provide interface designs and pathway structures that are more exploratory in nature, that appeal to a user's interest in playing with the medium. The success of such ventures, though, is dependent on the nature of the product and the willingness of the audience to play along.

Informational Programs

In addition to programming directly aimed at selling products, businesses and institutions also produce a large amount of more informational material. Interactive programs of this type include annual reports and a wide range of informational presentations designed to do such things as explain benefits, recruit new employees, or announce policy decisions.

Even though such material may have much in common with educational programs, their purpose is not so much to teach in a traditional sense as to convey a particular type of specialized information. Even within this context, though, designers need to be mindful of the company's image. In business, as in many public institutions, the company is always at some level selling itself—if not directly, then indirectly; if not to its customers, then to its employees and its board members, to public officials or the public at large.

Like advertising programs, informational programs often begin on an opening screen whose primary function is to distill the essence of the subject matter to a single image accompanied by a tag line. An opening screen, though, must also perform a navigational function. As such it usually presents the main category choices, or perhaps it provides a direct avenue to a navigational screen. On a project whose main purpose is informational, business program designers often eschew design metaphors altogether, opting instead for a categorical approach so that a design metaphor will not obscure a user's conception of the information.

The underlying structure of an informational program is often quite straightforward. Figure 8.4 presents a flowchart for a recruitment brochure designed on the behalf of a large telecommunications company seeking to hire recent college graduates. After the main menu screen, the chart divides into three main paths: Who We Are, Career Opportunities at Maximum, and Our National Network. This division reflects the designers' view that the audience, college students, is interested in three things: the nature of the employer, the jobs available, and the job sites. The goal of the design is to get users as quickly as possible to the information in which they are most interested.

Like interactive advertising brochures, informational programs take advantage of obligatory cuts and other interactive features to engage their audience. A recruitment disk, for example, might contain an application form that can be filled in, printed out, and mailed, or it might even provide a direct e-mail link to the company. An informational brochure whose goal is to explain the benefits of the company's pension plan might include a calculation tool that allows prospective employees to figure out their benefits upon retirement given different investments and anticipated rates of return.

One common type of informational piece is the institutional overview, which introduces the audience to the institution, its people, and its facilities. These programs are sometimes purely informational, as is a kiosk whose function is to familiarize patrons with the holdings and layout of a big city library. Most often, however, such informational pieces also have promotional elements. This is particularly true of interactive catalogues and guided tours produced by universities to help recruit students. A good example of one such program is an interactive tour produced by the Haas School of Business at the University of California at Berkeley. This CD-ROM opens with a series of obligatory cuts: a full-screen shot of the university campus, another of students congregating, and another of the tower at Sproul Plaza. This series of cuts culminates on the opening screen, which presents its navigational options by categories (see Figure 8.5). When the user moves the mouse to point to and click on one of the navigational buttons,

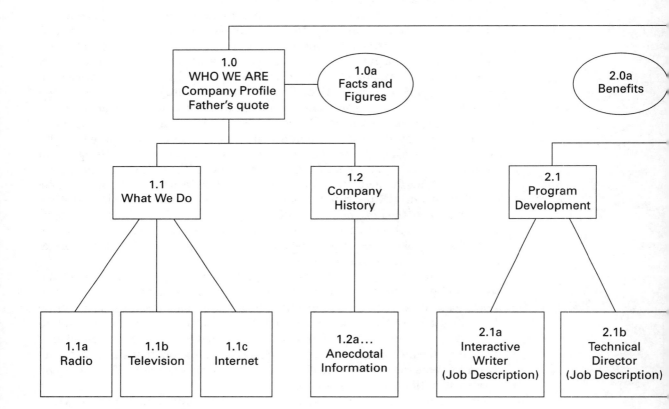

Figure 8.4
An Example of a Flowchart for an Informational Program. In interactive
business brochures such as the one charted here, broad, shallow structures
are often used to enable quick and easy access to information.

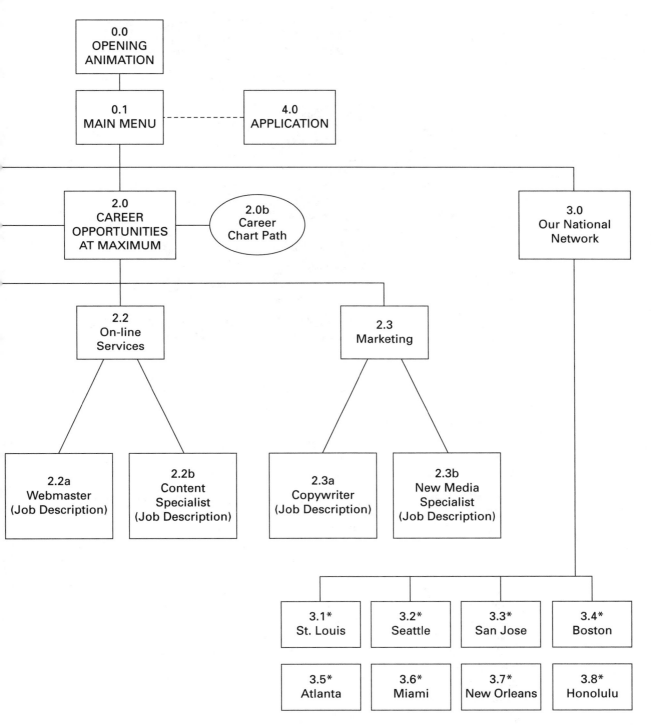

*Note: Items 3.1–3.8 are parallel items at the same submenu level.

Figure 8.5
An Opening
Screen with a
Classification
Scheme for
Navigational
Options.
(*Interactive Tour
of the Haas
School of Business,*
University of
California, Berkeley.)

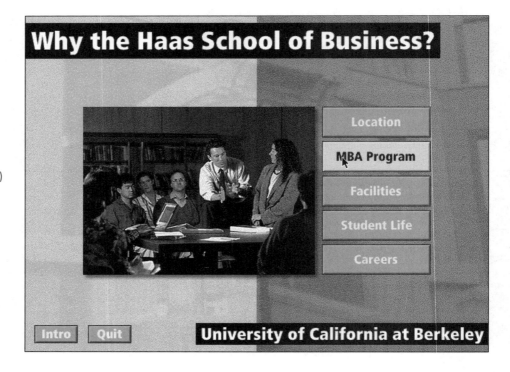

an audio narration of what's inside that particular section is activated. For example, if the arrow is over the button marked MBA PROGRAM, it triggers the following audio narration:

> AUDIO: This section includes information about the curriculum and faculty at Haas. It also details three special programs: International Business, Innovation and Entrepreneurship, and Management of Technology.

At this point the user has the choice of either entering the facilities module or staying on the main screen to explore other navigational options. The various subpaths are similarly organized using straightforward classification schemes, and a Quit button is always present. The user cannot exit without encountering two mandatory cuts. One of these displays the credits, and the other shows the students and faculty who helped with the project.

Interactive structures like these, based on easily discernible classification schemes, are common in informational projects, but they are by no means the only approach. Others rely on design metaphors that use no words and provide a far more game-like feel for users. One example is a portfolio piece by 3D JOE, which introduces a company named imedia to perspective clients (see Figure 8.6). In this program, the imedia office is rendered in 3-D, and users navigate around it by clicking on objects. Users can click on the imedia door to enter the office, then click on objects inside—the printer, the telephone, the television screen—to learn about the company's various services. Not all objects yield promotional information; some are there just for fun, as simple diversions. Users who click on a light socket will see sparks fly from the wall.

Figure 8.6
An Exploratory Interface for an Informational Brochure. Users who click on this door enter a virtual rendering of the imedia offices to learn more about the company and its services as they browse the environment. (© 3D JOE Corporation, www.3djoe.com)

Another example of a less categorical approach is *Smart Money*, a CD-ROM produced by San Francisco developers Adair & Armstrong for Beneficial Finance. This informational program takes the form of a game designed to teach high school students how to manage their money. In it, the players find themselves living in an apartment, with a monthly income and a charge account. The metaphor at the heart of the game is a no-holds-barred shopping spree, but there is a catch: If the students live too extravagantly, purchasing at will and spending beyond their means, they can sustain their lifestyle only for a while—eventually the bills come due and creditors come knocking. The purpose behind such a game, of course, is to teach students about managing money and the responsible use of credit. Companies such as Beneficial engage in the sponsorship of such programs as a form of indirect advertising, in the process enhancing their image in the community and winning future customers.

Training Programs

Increasingly, business programs are taking advantage of multimedia to create immersive environments that simulate on-the-job training. Such programs are built around the philosophy that the best way to learn is by doing. According to this way of thinking, training programs work best when they simulate real work, using real tools and allowing users to suffer real consequences for their actions.

Most traditional training courses begin with a particular instructional goal in mind, including a set of skills that need to be taught. Individual sessions are akin to modules, each of which is focused on a particular skill or technique. In a course on retail sales, for example, the instructor might begin each lesson by demonstrating certain sales techniques. The first session might focus on ways to approach customers after they have entered the store. The instructor might show a videotape that demonstrates both successful and unsuccessful sales approaches. Then, once the material had been presented, the next step in the instructional process would be to allow opportunities for practice. In this part of the class the students might get a chance to emulate the techniques they have seen demonstrated. They might try the techniques for themselves, perhaps practicing with their fellow students. After the practice exercises, the teacher would make comments and suggestions, reviewing again the main principles. At the end of the course, the students might undergo some type of evaluation—a trial run, perhaps —in which the instructor has a chance to critique their performance by watching them in real-life situations, perhaps in the store itself. The question of whether or not the course was a success, however, may not be answered until months later, when the boss ascertains whether sales have risen, fallen, or remained the same.

As in a traditional classroom, multimedia business training modules follow an instructional model similar to that discussed in Chapter 7: Introduction of Material. Presentation of Concepts. Practice. Review. Assessment.[2]

In another contemporary notion of training, rather than being shown first how to do something, students are instead thrust into the relevant environment and given a chance to take action. The emphasis is less on mastery of individual techniques and more on doing. The notion here is that students should do little passive watching; rather, it is better to give students the opportunity to fail, because that is when real learning occurs. In this approach, the instructor doesn't demonstrate something until a student asks to see it. Whatever information the teacher presents must be apropos of something the student just tried to do that didn't work out so well.

The first approach is more analytical, the second more experiential. Though both have their advocates, they need not be mutually exclusive, and most good programs make use of both teaching techniques. Still, most businesses are less concerned with educational philosophy than with the results of the training.

Business training programs tend to be of two general types. One is task-oriented and involves the acquisition of skills, including the ability to perform certain work routines. The other is more attitude- or behavior-oriented and focuses on employee morale and the ability to get along in the workplace.

Task-oriented training programs focus on concrete skills and definable work routines. Such programs might focus on simple tasks, such as training a new employee in the procedures for opening and closing the store, or they might be more complicated, involving the processing of a company's monthly payroll, the handling of plutonium at a nuclear site, or lab procedures at a university research facility. No matter the complexity, most of these tasks are routine at some level in that they are accomplished in certain stages that occur in a predictable order. In other words, they can be broken down into stages and analyzed as a process. Process analysis, as we have mentioned before, is one of the key organizational strategies interactive designers use when they adapt material to multimedia.

A flowchart for a task-oriented training program based on an instructional design model appears in Figure 8.7. In this program, the introductory sequence and

Figure 8.7
A Flowchart for
a Task-oriented
Training Program
Based on the
Instructional
Design Model.

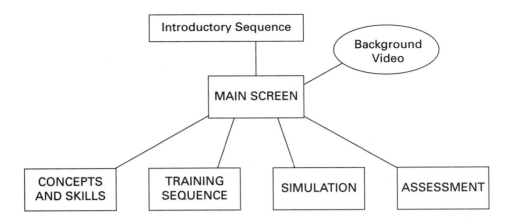

background video set the stage, appeal to the audience, and introduce the material. The main screen then provides access to the major parts of the instructional program. Moving from left to right in Figure 8.7, the first subscreen is devoted to introducing the foundational concepts for the rest of the program. The second subscreen is devoted to the training sequence, the particulars of which are charted in further detail in Figure 8.8. (In the training sequence, the process to be learned is broken down into stages. In Figure 8.8 it consists of five stages, each of which is followed by a practice or review session.) The third major subscreen on the main flowchart provides access to a simulation, which provides students opportunities for further practice and review. Finally, the fourth subscreen involves assessing how well students have learned the materials. Depending on the nature of the program, assessment might be purely for each student's own benefit, or the results might be tallied by the computer and made available to the instructor.

The structure described here should not be regarded as the only way to model a training program. Rather, it should be seen as a starting point and frame of reference. Not all training programs can be broken up quite so easily by process analysis and classification. This is particularly true of programs dealing with morale and employee behavior. In such cases, the goal is to improve the quality of human interaction. Although it is possible to teach fundamental concepts regarding behavior, sequential reenactment of those concepts in real-life situations is far more difficult. A conversation between employees, for example, does not necessarily move predictably from Point A to Point B. There are interruptions to contend with, immediate contexts to consider, long-term consequences to calculate—not to mention the conflicting goals of the people conversing. People differ in personality, mood, temperament, class, age, race, and gender, all of which affect human interactions.

As a way of simulating such variables during behavior training, some multimedia designers are incorporating game-like techniques involving simulated conversations between the trainee and on-screen characters. These conversations are structured and scripted in much the same way as games are. This is but another example of how writers and designers sometimes cross the traditional

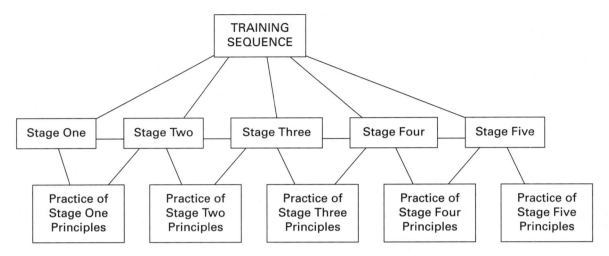

Figure 8.8
Detail of Pathways Connected to the Second Subscreen in Figure 8.7. This
portion of the flowchart concerns the five stages of the training sequence.

boundaries that separate education and entertainment, borrowing and adapting
ideas from one context and employing them in another.

The use of interactive media for training purposes has certain advantages. Students can work through the material at their own pace, returning to parts when
necessary and reviewing as much as needed. Multimedia not only provides visual
and audio aids to education, but also provides students with opportunities for
practicing skills in simulated environments in which they are free to make mistakes without real-life consequences. Interactive training programs can also be
used to give students instantaneous feedback on their performance. If training
programs are well-constructed, they can act as excellent supplements to real-life,
hands-on training and in some cases can even supplant it. (The decision to do the
latter, though, should be made quite judiciously, as there are limits to computer-
aided education.)

Training programs are one field in which design strategies for educational programs and business programs closely overlap. One of the challenges in writing
and designing for business is balancing the needs of the corporate culture with
those of the individual. In this regard, the writer/designer needs as firm an understanding of the audience as in any other medium, and an equally firm understanding of the instructional goals.

Format Considerations for Multimedia Business Programs

The format in which a business chooses to promote itself should reflect its clientele. There probably isn't too much reason for the corner grocery store to produce a multimedia CD-ROM. A colorful sign out front, together with coupons in

the local newspaper, might be a better way to reach its customers. A single principle holds true for all businesses: Each has to find the best media for reaching their particular audience.

High-end multimedia projects involving lots of video and audio and elaborate animations can be as expensive as a television show to produce, and they do not reach nearly as many people. Such ventures are likely to be reserved for high-end technology and media groups, who have large budgets and trade on their expertise in those fields. Even a big company, though, is unlikely to spend money unless some concrete advantage is perceived. High-end advertising projects requiring substantial investments of a company's time and money need to pay dividends, either by generating large amounts of publicity or by appealing to a very specific, well-heeled clientele. They can also be economical if produced in great quantity and distributed to a wide audience.

More realistic for most companies are smaller-scale projects that blend high-quality graphics and text. Over the past few years, universities and other institutions—particularly those trying to appeal to an audience with technical savvy—have increasingly placed informational catalogues on floppy disks. They have chosen this medium, as opposed to CD-ROM, because it is more economical, and they can reach a wider audience by choosing the more widely available format. Leading-edge technology may provide the flashiest media, but this advantage is lost if it relies on hardware that is not yet widely available. Moveover, it is possible to enhance the quality of floppy-disk presentations by putting the program on several disks and including an installer. The advantage of such an arrangement is that several disks hold more graphics than a stand-alone floppy and can even provide simple animations and brief sound bites. The disadvantage is they require users to run an installer program and shuffle disks in order to get the program onto their computer.

Then, of course, there is the Internet, which has become an increasingly popular format for businesses. Since its beginnings as a text-only medium for communication among researchers and scientists, the Internet's abilities to utilize graphics have increased and will likely increase further. Even as multimedia capabilities are added, though, producing works of quality will remain expensive. Therefore, a large number of Web sites will continue to rely on less-costly media. The absence of motion and sound, however, does not eliminate the need for design. When working on the Web, interactive designers must still address the same underlying concerns regarding audience, content, and structure that we have discussed throughout this text.

Many business are increasingly relying on Web sites as information directories, as places for potential employees, reporters, and investors to find information about the company. Toward this end, navigational structures are often straightforward, relying on category-based approaches. Similarly, the opening screens of Web sites tend to focus on establishing image, usually through the use of graphics and tag lines. As users navigate deeper into the program—to examine the company's annual report, for example, or to look for job openings or press releases—the colorful graphics may give way to an interface whose primary function is to present text. The balance between text and graphics, then, depends directly on a user's location within the program.

In another type of Web site, such as those promoting movies, text may be totally deemphasized: instead, graphic design and juxtaposition are highlighted. On some of these sites there is considerably more freedom in the way the interactive environment is structured, as the audience may be more willing to tolerate a meandering, associative structure akin to that found in computer games or interactive cinema.

Aside from building links within a Web site, another concern is how to link the site to others on the Internet. Of course, linking a Web site to other locations can be an efficient way of utilizing material that someone else has already placed on the Internet, or of calling attention to a related subject. The disadvantage is that the designer runs the risk of losing users altogether: When they go off to another site, they may never return. As a result, designers need to take care when establishing links to other sites. One way designers prevent user drift is by placing links to other sites on a separate screen that is off the main navigational pathways. Another strategy is to request that the remote site incorporate a return link. In this way, the links are still provided, but users are less likely to wander off before viewing what the designer wants them to see.

Finally, another program format for businesses involves Intranet systems. Intranet programs are similar to Internet programs in that both are transmitted via modem from a server to the client's terminal. The difference is that Intranet programs are not made available to the wider Internet audience—only the intended audience can view it. Aside from privacy, the principal advantage in Intranet systems is that they provide a high-bandwidth connection between the server and the terminal, thus making possible more rapid delivery of graphics and sound. Another advantage is that users are not at the mercy of the slowdowns, shutdowns, and information logjams that periodically occur on the Internet, particularly during peak hours.

The Design Process for Business Programs

In a business project, the design process begins with the first contact between the designer and the client. In an ideal world—at least from the perspective of the designer—such contacts would be initiated by a client who already has a project in mind. More likely, however, it is the other way around: The client needs to be contacted and sold on the project. The person who makes this contact and does the selling most often has a marketing background and is someone who understands the needs of the business. The marketing person in essence conducts a needs analysis (including a study of the audience), meets with the client, and then generates a proposal. In some instances the marketing person will even compose the initial flowchart and content outline.

The proposal is in essence a concept document covering major content areas, costs, delivery platforms, and format. It is likely to be fairly general concerning content and design strategies but more specific regarding the advantages the client might reap from such a project. The intention of the proposal is not merely to lay out the specifics of the program but to sell the project.

After the concept proposal is accepted, the client and the design team get together in a *kick-off meeting*. The primary purpose of kick-off sessions is research. The creative team gets a chance to understand better the client's needs, the audience, and the message that is to be communicated. The team also has a chance to plumb the resources of the company, to find out what content is already available, including photos, text, and graphics. The most important task, however, is to identify the message and tone the company wants to communicate.

If, for example, the client were a large telecommunications company hoping to recruit talented young people into its workforce, one concern might be to put a human face on the company. In other words, the client might want to downplay the company's large, corporate nature and instead emphasize its commitment to good working conditions and the surrounding community. In this case the message might center on a theme of bringing people together. Thus the writer and graphic designer, when looking through the preexisting corporate literature, will search for photos, graphics, and other content that contribute to this message. In addition, they will need to conceive classification schemes for organizing the program.

As previously mentioned, a good classification scheme for such a program might be similar to that charted in Figure 8.4. Based on such a flowchart, the writer would then produce a screen-by-screen treatment outlining the basic elements that will appear on the various screens. The point of this treatment is not to provide a template for production; that comes later. Rather, the intention is to create a working document that is the basis for further discussion and the refinement of ideas. Such a document might resemble the following treatment:

TREATMENT FOR EMPLOYEE RECRUITMENT PROGRAM

0.0 Opening Animation
The program begins with a brief animation, a layering of colors to form a stylized map of the United States, with villages and cities rising from vistas of green and blue. Intermittently spaced on the landscape are radio towers emitting transmissions and satellite dishes receiving them. The animation's atmosphere is folksy and down to earth yet incorporates technological imagery.

The logo builds on (appears on the screen, layering over the opening animation):

SCREEN TEXT:

> Maximum Communications:
> Bringing America Together

The three main navigational buttons, finely illustrated, also build on, as does the application icon.

0.1 Main Menu
Screen as above. Menu choices direct viewers into the following areas:

WHO WE ARE
CAREER OPPORTUNITIES AT MAXIMUM
OUR NATIONAL NETWORK

There is also both an APPLICATION icon, which directs prospective employees to a job application form, and a QUIT option.

This MAIN MENU is accessible from all first-level submenu screens.

1.0 Who We Are

This section features a halftone illustration of the company's founders, in their shirt sleeves, gathered in the company's first enterprise: a radio station in St. Louis, in 1938. A background montage shows the radio tower surrounded by citizens from various walks of life. The underlying intent here is to dramatize the idea that providing communications among everyday folks has always been at the heart of the company's business philosophy.

SCREEN TEXT:

In 1938, when my father started this company, his
goal was to bring the community closer together.
—John Williams

From this screen, users can delve deeper into the company's story by exploring one of the following categories: WHAT WE DO, COMPANY HISTORY, or FACTS AND FIGURES.

The user may also return to the MAIN MENU.

1.0a Facts and Figures

Text pop-up highlighting key facts about the company, with emphasis on its growth rate over the past decade and projected expansion onto the Internet. Intent here is to provide a brief view of Maximum Communications as a dynamic company and an engaging place to work.

Closing this pop-up returns user to WHO WE ARE.

1.1 What We Do

Screen display illustrating Maximum's main areas of corporate endeavor. This is an opportunity to show graphically what Maximum does, in a manner that blends photos and art work. Graphics should focus on radio, television, and Internet activities.

Embedded in the screen are three pop-ups offering further detail. One explores RADIO, the second TELEVISION, and the third INTERNET.

Other menu options here include COMPANY HISTORY and return to the MAIN MENU.

1.1a Radio

Pop-up text window highlighting Maximum's radio ventures. This consists of a concise list summarizing its key efforts in radio.

Closing this pop-up returns users to WHAT WE DO.

1.1b Television

Pop-up text window highlighting Maximum's television ventures. This consists of a concise list summarizing its key efforts in the television industry.

Closing this pop-up returns users to WHAT WE DO.

1.1c Internet

Pop-up text window highlighting Maximum's Internet ventures. This consists of a concise list summarizing its key efforts in Internet services.

Closing this pop-up returns users to WHAT WE DO.

1.2 Company History

Photo montage tracing the company's metamorphosis from a local radio station to a nationwide network of communications services, including cable television and broadband Internet services. This section is augmented by quotes—or key words—that focus on the company's values. The thematic intent here is to convey to prospective employees a sense of the company's history and values.

Other menu options here include WHAT WE DO and return to the MAIN MENU.

1.2a Anecdotal Information

Text windows, accessed from COMPANY HISTORY, that provide further information on the company. These are mostly anecdotal in nature and enhance applicants' understanding of Maximum Communications. Emphasis here is on the human aspects of the company's history, stories that impart a feeling for the company's unique place in the communities in which it operates.

Closing this pop-up returns user to COMPANY HISTORY.

The treatment itself is organized on a screen-by-screen basis. Notice that each screen segment is numbered to correspond to screens depicted on the flowchart in Figure 8.4. In addition, the treatment segment for each screen describes not only the various media elements that will appear on that screen but also the purpose and tone of those elements. Thus, under Screen 1.0, the writer not only describes the photograph but also mentions the purpose of the screen: to humanize the company and demonstrate its close relationship to the community. Such descriptions help provide cues to the graphic designers so they can better understand the tone they seek to communicate.

In addition to describing media elements, the writer also needs to communicate functionality for each screen. In other words, for any given screen the writer needs to indicate the navigational possibilities. The description of Screen 1.1, for example—concerning Maximum's primary business enterprises—details several navigational possibilities. First, it mentions embedded pop-ups that allow users to learn more about Maximum's endeavors in three areas: radio, television, and the Internet. These pop-up screens act as embellishments to the existing screen; the user calls them up, then closes them, without navigating to a new screen. On the other hand, the two other navigational choices here involve movement away from the current screen: one to the Company History screen, the other back to the Main Menu. Such information is important to the graphic designer, because it lets that person know what kind of navigational links and category buttons need to be present on screen. Also, it is vital to the programmer who eventually must build the software and construct the links between screens.

As mentioned, the treatment is not the final production script. Rather, it is a working document that will be submitted to the design team and to the client, both of whom will come back with additional ideas and suggestions. Based on those suggestions, changes will be made. The final production script will then include those changes, as well as precise descriptions of the exact pictures and media elements to be used.

SUMMARY

The design process for interactive business programs has much in common with the design process in other types of interactive media. The difference rests in the audience, and in the two-way communication between the design team and the client. The writer's role is often that of compiler and reviewer of information, arranging and classifying and then synthesizing a message that is appropriate in tone and content for the audience. During this process, there is input from two sources: the client, who will want to review the script at various stages to be sure that the message suits the company's purposes, and the other members of the design team, who will need some creative flexibility when it comes to implementing the design because ideas in script format do not always translate perfectly to the screen.

In business applications, as in other fields, a person with good writing and communications skills can have a profound impact on the shape of a media project. Business programs are in fact particularly reliant on written scripts, partly because of clients' involvement in the process and their desire to control the message. Even though the computer-game industry might still turn to programmers to develop content, this is rarely true in the business world, which is much more reliant on creative professionals with backgrounds in communications and marketing.

EXERCISES

1. Choose a company or product that interests you: a tool maker, a record company, a bookstore, or a new line of clothes, for example. Then collect as much information as possible about that company or product, including any written or graphic material such as brochures, catalogues, and so on. After examining your material closely, go through the following six-part process to create an advertising program for interactive media.

    a. Determine the audience and message of the company's existing literature.

    b. Determine the format that might be best suited for adapting the message to interactive media.

    c. Conceive of a design metaphor or classification scheme for your project.

    d. Create a concept proposal for this project. In this proposal describe the audience, the message, and the main content areas.

    e. Draw a flowchart that maps out the project.

    f. Write a screen-by-screen treatment of the project.

2. Go through the process outlined in Exercise 1 to create an informational program for interactive media.

3. Analyze an existing training program and develop a strategy for adapting that program to interactive media. First, do a process analysis to identify each step in the training program. Then draw a flowchart that breaks that process into its main stages. Finally, write a paragraph for each part of the flowchart describing what key concept needs to be communicated on each screen.

ENDNOTES

1. William L. Rivers and Alison R. Work, *Writing for the Media* (Mountain View, CA: Mayfield Publishing, 1988), 213–234.

2. For information regarding informational and training programs, see the following: Walter Dick and Lou Carey, *The Systematic Design of Instruction* (Glenview, IL: Scott, Foresman and Company, 1985); R. Kaufman and F. W. English, *Needs Assessment: Concept and Application* (Englewood Cliffs, NJ: Educational Technology Publications, 1979); R. M. Gagne, *Conditions of Learning* (New York: Holt, Rinehart & Winston, 1977); D. A. Payne, *The Assessment of Learning: Cognitive and Affective* (Lexington, MA: Heath, 1974); L. J. Briggs, *Handbook for the Design of Instruction* (Pittsburgh: American Institutes of Research, 1970), 93–162; M. Cagne, W. Wager, and A. Rojas, "Planning and Authoring Computer Assisted Instruction Lessons," *Educational Technology* 21, no. 9 (1981): 17–26; R. A. Reiser and R. M. Cagne, *Selecting Media for Instruction* (Englewood Cliffs, NJ: Educational Technology Publications, 1983); J. D. Russell, *Modular Instruction* (Minneapolis: Burgess Publishing Co., 1974); and J. P. DeCecco, *The Psychology of Learning and Instruction: Educational Psychology* (Englewood Cliffs, NJ: Prentice-Hall, 1968), 54–82.

9

Technical Considerations in Multimedia Projects

Interactive media, more than other media, demands that writers have a technical understanding of its inner workings. This is so partly because the technology itself has not yet matured, and writers and designers need to be aware of constantly shifting capabilities. Such a situation has its inherent dangers. If writer/designers give the technical considerations too little attention, they risk creating programs that simply fail to deliver content in the intended ways. On the other hand, if technical considerations are given primacy—as is too often the case—then designers run the risk of overwhelming the content in a search for technical innovation. The work of the writer, then, is to regard technology not as an end in itself, but instead to seek a careful balance in which the technology serves the content and—most importantly—the audience for which that content is intended.

This chapter, which focuses on technical considerations of interest to writers working in interactive media, begins with a discussion of the relationship between conceptual and technical skills. It then discusses methods for working within the technical limits and concludes with a section on software tools for multimedia writers.

The Relation of Conceptual Skills to Technical Issues

In a book about technology-driven media, it may seem odd to suggest that a writer's best and most useful tools are still the oldest ones. Nonetheless, this is true. It is more important for a writer/designer to have a good grasp on traditional rhetoric, the ancient art of persuasion, and the craft of storytelling than it is to be a master of the latest programming language or graphics software, or of any of the other countless software packages that emerge with each turn of the computer industry's product cycle.

One underlying contention of this text is that the basic principles of design we have discussed will continue to be valid no matter the direction in which technology evolves. This contention seems borne out by the fact that much of the writing and design work we have discussed does not require elaborate software skills. The conceptual process can be accomplished with the oldest of implements because the fundamental conceptual work is based, by and large, on principles that are much older than today's technology.

Flowcharts, for example, can just as well be drawn by hand as with specialty software. During the scripting of *In the 1st Degree*, the charts were drawn with felt markers on a huge white board. This arrangement allowed for the simultaneous physical participation in the process of three people, each with a marker in one hand and an eraser in the other. The head writer created flowcharts at home using pencil and paper and wrote much of the dialogue and story treatment out longhand before entering it into the computer.

Still, it would be a disservice to suggest that interactive writers should not be familiar with certain software tools or with the fundamental technical issues that affect production. Some production teams require a very high level of technical literacy from their writers. Even when this is not the case, a time almost always comes when the conceptual information must be entered and arranged on the computer. Producers like to have their scripts on computers because it makes them easier to edit and distribute. Media professionals have grown used to the high-quality type and graphics that printers produce. Moreover, there are other advantages to working directly on the computer, particularly when the end product requires the integration of text with graphics and programming code.

There are also some very pragmatic reasons for being conversant with technological issues, particularly the capabilities of multimedia software and hardware. Such knowledge can help writers understand the limits of the media for which they are writing. This is of particular importance in a developing medium such as multimedia, for which the boundaries are constantly changing. In this regard, an understanding of some of the underlying technical issues is vital. The first such issue we will examine involves platforms and formats.

Platforms and Formats

In interactive media, questions regarding platform and format always arise. In general, *platform* refers to the hardware, the machine on which programs are run. *Format*, in contrast, refers to the software that acts as the delivery vehicle for programs.

IBM PCs and Macintoshes are examples of two different platforms. Floppy disks, CDs, game cartridges, and laser disks are examples of different types of formats. The choice of format and platform determines in part what is possible on any given project, including the availability of graphics, sound, and video.

Questions concerning format and platform are complicated by the fact that the possibilities are fluid, existing along a continuum that is in constant flux. Part of this flux stems from the technology itself. Not long ago, it was difficult to put four-color imagery on a home computer, let alone simple animation; now both of these capabilities are relatively routine. As compression rates increase and bandwidth problems are resolved, new formats and more powerful platforms are likely to develop, offering more possibilities.

Given that change is a constant, how is a writer to keep up? Is it necessary to know each medium's precise storage capacity, measured in bytes and RAM, and how these capacities are affected by the choice of media elements, programming language, and compression schemes? From a writer's point of view, such in-depth technical knowledge is not often necessary, unless the writer is also the technical producer. In most circumstances it is more helpful to think in terms of the continuum of format and platform possibilities. At the near end of the continuum are disk-based programs containing text and a few simple graphics; at the far end are the three-dimensional, virtual worlds of the future. Somewhere in-between are CD-ROMs and the current capabilities of the Internet.

For our purposes, the first step along the continuum is the common floppy disk. A great deal of interactive material—for entertainment, education, and business—has already been produced for floppy disk. Such programs tend to combine words and still pictures with only minimal sound or animation. The hallmark of such presentations is that they are designed to operate on the most commonly available technology, so that they reach as wide an audience as possible. Not too long ago, very few computers had CD-ROM players, and many still do not. The floppy disk was (and still is) a format capable of reaching a wide audience of computer users. What it gains in audience accessibility, though, it loses in presentation capability. Floppy disks simply do not have much storage capacity.

The next step along the continuum is represented by formats that allow users access to more graphics and to some sound and motion but still fall short of full multimedia capabilities. These formats include some specialty disks and programs stored on local servers. The Internet itself is a good example of a presentation format that goes beyond the floppy disk but does not offer full multimedia capabilities (at least not as of this writing). Because of its vast storage capacities, the Internet allows users to browse far more color graphics than could ever be stored locally. It also provides downloadable animation, video, and sound, the possible uses of which depend on each user's machinery and technical skills. In addition, new Web-browsing tools and other applications promise to extend the Web's multimedia capabilities yet further. Software add-ons, such as *RealAudio* and *Shockwave*, are making it possible for users to view more sophisticated animation without downloading and manipulating files.

Another step along the continuum involves programs that include moving pictures and CD-quality audio. Some of the first ventures into this capability came from game developers using video cartridges in tandem with home computers and television monitors. Another early variation was the stand-alone multimedia kiosk, in which the storage medium for audio and video material was a laser disk. More recently, developers have come up with more powerful compact-disk formats, the most well known of which is CD-ROM. These formats allow for increased storage and playback of video and audio files, but they still fall short of the quality that is commonplace in television and film.

The great enthusiasm for CD-ROMs in the early 1990s has dampened as consumers have shied away, complaining of compatibility problems, slowness of operation, and a lack of high-quality content. Attention then turned to the Internet, where new software developments have increased multimedia capabilities. Interest continues to move back and forth between these two formats; some developers are using them in tandem. A new digital disk format forthcoming, known as *DVD* (Digital Video Disk), may eventually replace CD-ROM, as it can be played on a television set and offers greater storage capacity at higher resolution. Whether it works as advertised, and whether consumers are in fact interested, will largely determine its viability.

Current Technical Limitations

Throughout the early years of multimedia, developers have struggled with the task of bringing clear, crisp video to the computer screen. The task has not been easy. For CD-ROM titles, the process has involved digitizing the video and then compressing it so it can be stored in discrete files on the disk. To enable viewers to play visual sequences on their home computers, these video files must be coded and arranged by programmers and then packaged with playback software. Though this process has become relatively routine in the past few years, frequent glitches still occur. Playback software still crashes more than anyone would like, freezing the screen and leaving the user with no choice but to reboot. Load times are slow, video quality falls short of television, and restrictions exist on the amount of video available and the size and quality of the image.

Part of the struggle involves the nature of video itself. Broadcast-quality video of the sort we see on television is recorded at thirty frames per second. This means that when we are watching video, thirty individual pictures flicker by each second, which is about the right speed to create the illusion of real-time motion. In order to capture this experience on the computer, each frame—with its many thousands of pixels—must first be digitized and stored in memory. Because each second of video consists of thirty such frames, the storage requirements for even a few minutes of video are enormous, equivalent to those of many, many thousands of words of text. Thus, for video, disk space is always at a premium, even given the relatively large storage capacity of CD-ROM.

To save disk space, developers and programmers have developed a number of techniques for handling video. One such technique involves the use of compression schemes such as MPEG, which "shrink" the video so that it takes up less space. Compression schemes are based on the assumption that much of the information stored on consecutive frames in a piece of video is actually repetitive, and that the repetitive information can be removed. For example, in a five-second clip showing a sprinter, the runner's body might be moving as she hustles straight down the track directly toward the viewer, but much of the background remains unchanged: the sky, the mountains extending above the top of the stadium, the dirt track at her feet, the color of her shirt. Compression schemes operate by remembering the information that repeats from frame to frame and throwing out duplicate information. If for each frame the computer can reuse the pixels for the sky and replaces only the image of the woman as she moves down the track, then it is possible to save disk space.[1]

Another way to save space on the disk is to limit the size of the playback image. An image that occupies only one-quarter or one-eighth of a screen takes up far less storage space than does full-screen video. Similarly, playback quality is limited by the capabilities of QuickTime, the software used to play back most computer video (see Figure 9.1). As opposed to the conventional thirty frames per second, QuickTime drops every other frame, running at only sixteen frames per second. This reduced frame rate produces an image that is pixelated, a bit herky-jerky, and frequently out of sync with the sound. Altogether, these limitations can produce an effect not unlike that of watching an old-time, poorly dubbed foreign movie on an underpowered projector.

Not only do space considerations affect the amount of video available; they also affect quality. The gap between traditional, broadcast-quality video and CD-ROM video has narrowed somewhat recently, partly because of improvements in QuickTime but also because developers have become more adept at disguising the flaws. On early CD-ROMs, the video was displayed in windows not much larger than postage stamps and was often grainy as well. More recently developers have adapted blue-screen techniques from television (see Figure 9.2), in which actors are videotaped in front of a blue but otherwise empty background. Then the actor's image is effectively cut out and removed from the original back-

Figure 9.1
A QuickTime Video Window. On early CD-ROMs, compression techniques limited video display to small windows on the computer screen. (*Macbeth*, A. R. Braunmuller and David S. Rodes, eds., The Voyager Company.)

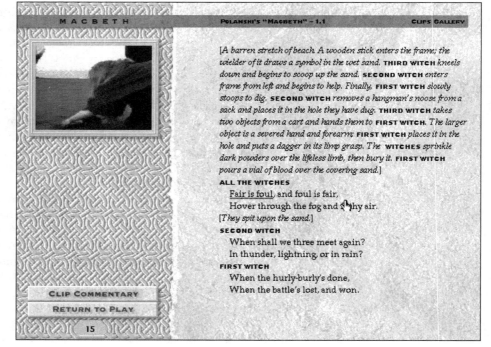

Figure 9.2
Superimposing
Actors on a Digital
Stage. The use of
blue-screen
techniques,
pictured here,
enables producers
to mimic full-
screen video by
superimposing
actors onto digital
backgrounds.
(*The Making of
In the 1st Degree*,
Broderbund.)

drop and superimposed onto a new background. This technique not only allows for dramatic use of backdrop but also helps compression schemes run more effectively and limits the jerkiness of QuickTime. The technique is similar to that used during the weather report on evening news shows throughout the country, in which the audience sees the weather person standing in front of a digitized weather map one moment, then some beautiful scenery the next. The location of the speaker, of course, is not changing; instead, the background image is being recomposited electronically in the studio.[2]

On the Internet, the problem has not so much involved storage space as delivery. In theory, large, distant servers can hold vast amounts of video and audio clips, which then can be dialed up via modem and played on home computers. One difficulty, though, rests in the limited transmission capabilities of the phone lines that make up the primary connection between remote Internet locations and the home: The current wiring systems do not have enough bandwidth to transmit video and audio effectively. Until recently, multimedia developers working on the Internet have been limited to the use of still graphics and text; audio and video were present, if at all, in files that users could download and view off-line. This situation obviously required dedicated viewers with patience, the proper equipment, and a certain level of computer skills not possessed by the average user. Furthermore, download time created a severe gap in program continuity. Even though rapidly developing applications such as *Java*, *Shockwave*, and *RealAudio* are now bringing more immediate multimedia capabilities to Internet users, these technologies are still in their infancies and lack the quality of film and television. Moreover, delivery is often slow and quirky, particularly to home users relying on conventional phone lines and modems.

Despite the limitations of the World Wide Web, there are other ways to deliver multimedia over networks. The secret is in expanding the transmission capabilities of the wire that links the server to the user. Traditional phone lines seem to lack that capability, but other wiring systems have larger bandwidths. ISDN, one such system, is available to home users who are willing to pay for it. T1 and T2 lines provide faster options, but the expense of installation generally limits them to use by large institutions in Intranet systems. Indeed, used in conjunction with cable modems, any of these systems can deliver multimedia, but their use has been limited because of the cost of installing them. Meanwhile, many developers are focusing on developing techniques for encoding video and audio signals so they can be streamed over traditional phone lines. Although these improvements seem promising, phone companies are concerned that sooner or later the expanding use of the Internet for such purposes will overburden the phone system, interfering with the companies' ability to deliver traditional phone service.

All these delivery systems are imperfect; all of them are also relatively new and thus likely to undergo further change. Many developers contend that it is difficult to reach an audience without a standardized delivery medium, let alone predictable hardware. The question for now revolves around the delivery system and whether or not the primary vehicle for multimedia will be some enhanced version of CD-ROM, the Internet, or even the television set.

Regardless of which technology dominates, we can be sure that before this textbook has gone out of print (or perhaps before it even reaches print), advanced compression techniques will enable enhanced multimedia capabilities for an increasingly wider audience. The far end of the continuum will continue to provide new possibilities. Off in the future, somewhere, are more fully immersive environments foreshadowed by current experimentation with virtual reality headsets, giant wall screens, and electrochemical impulse sensors implanted in the skin.

Working Within the Technical Limits

Given that the technology is always changing, the task of writing and designing within multimedia system parameters may at times seem all but impossible. However, it is really not as difficult as it seems. A given project will always have limitations that dictate a certain mix of audio, visual, and textual elements. As a result, each project will fall somewhere along the continuum we have been discussing.

The technical details of media production are not usually the primary considerations of the writer but are handled by the producer or technical director. Sometimes writers are not even brought into a project until basic format and technical decisions have been made, in which case a producer may dictate the length and type of writing that needs to be done. This is not always true, however, particularly if the writer also shares producing duties or is otherwise involved from the project's inception.

Regardless, some basic technical considerations affect the work of the writer: the relative proportion of media elements, format capabilities, computer memory, and software capabilities. We will discuss these topics next, as well as some strategies for working with other team members in expanding the technical limits.

IDENTIFYING THE PROPORTIONS OF MEDIA ELEMENTS

The technical considerations of perhaps the greatest concern to writers are those that affect the availability and relative proportions of the various media elements. Put simply, writers need to know how much audio, video, and graphics are available. Another major concern is the creative capabilities of the programmer, both in terms of staging special effects and building contingencies into pathway construction. All these concerns affect every stage of the writing and design process.

For example, a high-end CD-ROM project produced in the early 1990s typically included roughly one hour of video or animation, maybe several dozen full-screen graphics, and perhaps two hours of audio. It could also hold many millions of characters of text. The exact proportions of the various elements varied for different projects, but there was one general rule: The less video, the more of everything else. This fundamental observation still holds true today, even for DVD. In contrast, developers on the World Wide Web during the early 1990s were restricted to the use of graphics and text. Graphics capabilities have exploded in the last few years, and it is now possible to use some video and audio on the Web. Still, the difference in capabilities between formats can have a considerable effect on the writer's approach to the design process at all stages, from initial proposal to the final production script.

UNDERSTANDING FORMAT CAPABILITIES

To get a better understanding of how format capabilities can impact design, consider the following example. Suppose that a writer is crafting a proposal for an educational program designed to teach high school students about the weather. In the proposal stage, the writer needs to know the upper limits of the format's capacity because these limits affect everything, even the nature of the research. A proposal for CD-ROM, for example, might feature a guide figure, a well-known climatologist who appears on-screen and informs viewers about the fundamentals of weather prediction. It might also propose the use of video clips to show catastrophic weather events, such as tornadoes and hurricanes. The program might even include a section that allows students to input variables concerning a given weather system—changing wind speed, air temperature, and atmospheric pressure—and then track the results as the weather system moves across the landscape.

On the other hand, for a similar project on the Web, a proposal writer might first need to consider whether this project was going to be delivered over T1 lines using a dedicated server or instead was to be transmitted over a narrower bandwidth. Even at the broadest bandwidths, the designer might eschew the kind of motion- and sound-intensive features mentioned above and instead focus on still graphics and text material and use video and animations in a more limited way. Still, many types of branching possibilities are available to the Web designer, including the possibility of linking to other Internet sites and the ability to supply chat forums and real-time weather information from around the globe.

Similar differences apply to the rest of the design process. In the treatment stage, the CD-ROM or DVD writer will seek a design metaphor and conceive an interface that will accommodate the use of video. The placement of video and audio in the navigation paths—where these elements will be used, how often, and in what amounts—are major considerations. The Internet writer, on the other hand, has a different concern. Because of bandwidth considerations, video and sound are often more ornamental in nature and may not be used at all. The focus traditionally has been on ways to balance text and image in the context of the interactive structure.

The nature of production script, too, varies according to the format. All production scripts need to address programming concerns and the relationship of each element to the others. In CD-ROM and DVD projects, video and audio items on a script need to be tagged, labeled, and cross-referenced to a flowchart for use by the programmer when assembling the various elements. Moreover, a writer on a media-intensive project needs to be cognizant of the relative space available. If the producer has budgeted only a half-hour of interactive video, then the length of the script must be adjusted accordingly. (A general rule in video is one page of formatted, typewritten script equals one minute of playback time. Because script formats vary so much in multimedia, however, this can still be tricky. The best way to determine the length of a given scene is to time it while speaking the lines out loud.) Until recently, many of these considerations have been secondary on the Web; designers have been concerned instead with the load times of still graphics. This remains the case for many Web projects. In the production script, the writer often focuses on the screen text itself, crafting it to match the unique requirements of the Web.

Whether working for the Web or disk, though, writers also need to be cognizant of screen size when specifying animations or graphic materials. It is essential to keep in mind the limited size of the computer screen and the fact that screens must often accommodate a number of different kinds of visual information simultaneously, including icons, buttons, pull-down menus, and navigational tools. Similarly, another concern in almost all interactive projects involves working within the limits of available computer memory.

CONSERVING COMPUTER MEMORY

In almost any project—particularly when video and animation are involved—bandwidth and disk space are important considerations. (On the Web, the issue may not be disk space but file size and its relationship to download time.) In this regard, writers can use several techniques for saving memory space.

The first has to do with the efficient structuring of interactive pathways. One obvious approach involves reserving the high-end media glitz for the most well-traveled pathways. Because full-motion segments are usually the most expensive to create and the most memory-intensive, it is generally a good idea to place them where users will likely see them, not on seldom-traveled, tertiary pathways. On the Web, however, where not all users have equal bandwidth, the opposite can be true. Instead, designers may have to arrange multimedia segments in ways that allows users to bypass them easily because transmission over ordinary phone lines can be unwieldy and time-consuming.

Another technique for conserving media resources involves the effective use of audio. Audio does not take up nearly as much computer memory as video. When audio is dubbed over a slide show presentation consisting of rapid dissolves from one slide to the next, the effect can mimic that of full-motion video or animation. Sometimes, in fact, the effect is superior, because still-motion imagery can be used to fill the entire screen, whereas QuickTime video is limited to a small window. High budgets and the latest compression technologies do not always create the best effects.

UNDERSTANDING SOFTWARE CAPABILITIES

In addition, writers need to be familiar with the capabilities of programming software, which can affect not only the media used but the interactive structure as well. Most branching structures for most interactive programs are fixed and immutable; to borrow a term from the programmer's lexicon, they are *hardwired*. This means that the relationships between the user's actions and the computer's response are predictable and fixed. When the user pushes Button A, the computer initiates Response A; when the user pushes Button B, the computer initiates Response B; and so on.

Sometimes, however, the branching structure needs to be more complex and more responsive to different situations. In such instances, the programmer can take advantages of the computer's ability to keep track of many variables at once, not just the user's immediate position in the path. This interactive structure is called *conditional linking*. If, for example, the computer knows that a user has already pushed Button B before, this time it responds with Response C, or, in a different circumstance, with Response D. Such capabilities allow writers to add a great deal of complexity to narrative pathways by changing the computer's behavior according to both the short-term and long-term actions of the user.

The best way to understand the possibilities of a particular format is to experience programs created for that format. To understand what is possible on the Web, writers need to search for, read, and navigate various Web sites. The same holds true for CD-ROM programs or for any other format. After designers have studied a large number of programs, it is easier for them to see how a project's possibilities are circumscribed not only by technical considerations, but by its budget, its purposes, and its audience, and even by the talent of the creative people involved. Once such things are understood, it is not only easier to understand the limitations of a particular project, but also how those limitations might be overcome and the possibilities expanded.

EXPANDING THE TECHNICAL LIMITS

Aside from examining what has *already been done*, writers need to expand the boundaries of what *can be done*. More than just a general understanding of technical issues, this takes imagination. As much as anything else, though, it takes a willingness to talk to others on the design team, particularly the graphic designers and programmers, for talking about new possibilities is the first step toward implementing them. Moreover, one way to gain the respect of programmers and graphic designers is to have some understanding of their jobs and the creative challenges they face.

The leading edge of the technical continuum will continue to advance; as time goes on, more will become possible. This does not mean, though, that writers will suddenly find themselves without restrictions. Practically speaking, in the future as now, writers will find themselves limited not only by format but by budget. As always, writers will find themselves working within certain constraints.

The important thing is to know what is possible in a given format. At the same time, it is important to remember that high-end technical capabilities are not necessarily synonymous with good work. Everyone knows that a five-minute documentary video can be more moving than a big-budget Hollywood movie, just as a well-written paragraph can convey more meaning than an entire volume of badly rendered prose.

The same holds true in multimedia. To be successful, a project need not be high-budget or rich in technical pizzazz. Rather, writers need to be aware of where their projects fall along the continuum because of the general limitations and possibilities of the formats in which they are working. They do not necessarily need deep technical knowledge. Multimedia writers, for example, do not need to know the specifics of compression schemes or the amount of RAM taken up by a given audio file, any more than screenwriters need to know the specifications of the electrical generators used to power the klieg lights on a location shoot.

Still, writers do need to understand that they are working in a technical arena, and it is helpful to speak the language and understand the concerns of those in charge of technical implementation. Such knowledge can be helpful, not only in understanding the dimensions of the media, but also in helping to extend the continuum itself.

Software Tools for Multimedia Writers

Two software tools are especially useful to writers. One is a good word-processing program; the other is a charting program for creating flowcharts. Even though some of the major word-processing programs have recently begun to build charting capabilities and hyperlinking functions into their packages, these programs are somewhat clumsy and difficult to use.

Several stand-alone charting programs can help writers construct flowcharts similar to those used in this textbook. One such program is called *Top Down;* another is *Inspiration.* Both of them are relatively easy to use but have only limited word-processing capabilities. For that reason, some writers prefer specialty programs such as *Storyvision* or *Script Thing,* which are compatible with word-processing programs and facilitate written composition while working within a chart. However, these programs also have their limitations.

In addition to having charting and word-processing programs, writers often find it necessary to expand their software repertoire in response to the needs of a particular project.[3] Sometimes, for example, a writer/designer might find it necessary to build a working prototype that more closely resembles the final product, including, perhaps, some rudimentary graphics. Along these lines, Web writers have found it helpful to learn *HTML*, or hypertext mark-up language, a program that allows them to make screen breaks and insert graphic placeholders, thus creating prototype versions of Web sites.

Similarly, *HyperCard* is a relatively easy-to-learn prototyping tool that can be used to test out both educational and entertainment projects. It allows the user to build screens, construct links, and even run animations. *HyperCard* has been used as the primary operating code for a number of commercially successful projects. A similar software tool, called *HyperStudio*, is currently finding popularity as an authoring device. With either of these programs, though, one should keep in mind that there is a big difference, in both form and function, between rudimentary prototypes and full-fledged programs that have been professionally enhanced with sound and animation.

Another common software tool used in multimedia applications is *Macromedia Director*, a programming language reputedly designed for artists and creative types who do not want to get into the intricacies of advanced programming languages. However, although *Macromedia Director* may offer some advantages to less technically oriented writers, it can still be a difficult program to master. How difficult is a matter of debate. Some developers stand behind *Macromedia Director* as the best tool for developing CD-ROM applications, particularly in entertainment projects. Others say that it is ultimately easier to program in *C* (or in one of the other more-advanced programming languages) than to deal with the idiosyncrasies and limitations of *Director*.

Similarly, developers of educational projects use a programming tool known as *Authorware*. For educational programs, *Authorware* offers certain advantages over *Macromedia Director*, primarily its ability to incorporate multiple choice questions and other testing functions. For this reason, *Authorware* is sometimes preferred by producers with an instructional design background. Again, however, some users say that *Authorware*, like *Macromedia Director*, can be clumsy to use compared to the more elegant and powerful programming languages upon which they are based.

How far a writer/designer wants to go in learning how to program is an open question. Certainly it is possible to develop programming skills, just as it is possible to learn how to use *Photoshop*, *Adobe Illustrator*, and various other graphics programs. How much one wants to explore these other, more technical skills is a matter of personality and natural proclivities. Given the current emphasis on technical skills and cross-disciplinary learning, the ability to work across boundaries seems advantageous. Still, each of us has to be concerned about diluting our talents and skills and must remember that all the technical ability in the world will not compensate for the lack of strong conceptual skills. The temptations are strong: Software developers are always coming up with new programs that are supposed to make life easier in one way or another. The same is true regarding the various tasks of interactive design, whether it be for educational, business, or entertainment programs.

When it comes to instructional design, a program named *Designer's Edge* walks designers through the instructional design process, helping them to create training programs. *Designer's Edge* divides the design process into its component parts, helping with the development of goals, concept documents, objectives, and, ultimately, the instructional course itself. The purpose of the program is to provide a template for computer-based education, with particular emphasis on training courses for business.

In the realm of games, programs such as *Inform* and *Adventure Simulator* can be used to create text-based adventure games. Similarly, Eastgate Systems offers a program called *Storyspace* designed for interactive fiction writers. Well-known game designer Chris Crawford has reportedly been working on a scripting program for writers who work in interactive cinema and computer games.

At their best, such programming tools are great aids to creativity, providing a structure and a venue in which to work; at their worst they can be stifling. The trouble with such tools, particularly if they impose organizational constraints, is that the programs they help to create tend to be much the same regardless of the designer. They tend to have the same conceptual approach and the same overall structure. This might be all right in certain contexts, but such a formulaic approach can also be deadly if overused. In that case it may be beneficial to forsake the software and return to pencil and paper to burst the boundaries the program creates.

Software programs have limitations, and sometimes they can even act as an obstruction to creative vision. No design program can make you a good designer, anymore than a word-processing program can make you a good writer. They are tools, and it is up to you to figure out when and how to best use them, and when judiciously to place them aside.

S U M M A R Y

The primary skills that a writer brings to any media project are conceptual skills: the ability to conceive story lines and patterns of exposition, to organize and classify information, to develop plot and character, and to write dialogue and on-screen text. In interactive media, writers need not only these skills, but also a deep understanding of the complexities of interactive design. In addition, writers need to be aware of the technical limitations that constrain possibilities within the media.

Writers need to be familiar with the various platforms used in interactive media, as well as the general capabilities of the various media formats, ranging from simple floppy disks, to CDs, to specialized programming delivered over networks from remote servers. Because the technology is constantly evolving, this can be a difficult task. Therefore, it can be useful to think in terms of a continuum arranged according to the types and proportions of media elements generally used in different projects. Rather than developing specific in-depth technical expertise, it is usually more important for writers to be generally conversant with a wide range of technical and graphic matters, so that they can communicate with visual and programming artists, producers, and technical directors about what is possible within a given format. The writer's primary interest is not in technology for its own sake, but in how technology affects the relative proportions of the media elements that are an integral part of the conceptual design.

The primary software tools used by most writers are simple charting and word-processing programs. The charting programs are useful for visualizing interactive structure and help impart a simple, graphic dimension to structures that are difficult to describe verbally. They are also useful in helping to conceptualize the screen-to-screen movement of a given program. Sometimes, though, more detailed rendering is needed, a function provided by a number of programs, ranging from *HyperCard*, to *Macromedia Director*, to custom programming written in *C*, *C+*, or other programming languages.

Given the wide range of skills needed in multimedia projects, familiarity with the mechanics of prototyping, screen design, cinematography, and even programming can be of great value to a writer. How far writers should go in developing these other skills, though, is a question of personality, individual circumstance, and interest. Even though it can be tempting to immerse oneself in an ocean of technical details, such immersion is not always relevant to the larger, more crucial questions of conceptual design. Most producers rely on writers for their conceptual skills regarding the meaning, depth, and arrangement of content. If matters of content are not addressed fully and considered in terms of the intended audience, then the project is likely doomed to failure: All the technical expertise in the world cannot rescue a poor conceptual design.

EXERCISES

1. Working in a group, choose an important social or political event of the past year and analyze it. What were its beginning, its middle, and its end? Who were the participants? Where did it happen? How did it affect the participants? How did it affect other people? What were its long-range consequences?

2. Compile your answers for Exercise 1 into a content document.

3. Using the subject matter generated in Exercises 1 and 2, construct a flowchart and a concept treatment for an interactive program on floppy disk.

4. Using the same subject matter, construct a flowchart and a concept treatment for an interactive program on the Internet.

5. Once again, use the same subject matter to construct a flowchart and a concept treatment for a program on CD-ROM or DVD.

ENDNOTES

1. Ken Fromm and Nathan Shedroff, *Demystifying Multimedia* (San Francisco: Vivid Publishing, 1993), 160, 280.
2. Michael D. Korolenko, *Writing for Multimedia* (Belmont, CA: Integrated Media Group, 1996), 135.
3. Darcy DeNucci, ed., *The Macintosh Bible* (Berkeley, CA: Peachpit Press, 1994), 671–726.

10

Interactive Media and the Future

A book on interactive media almost has to end with a glance toward the future. In a medium that is so new, in which so much is undecided, and so much likely to change, everyone wonders what lies ahead. The trouble is that predicting the future is a tricky endeavor, prone to all types of errors, not the least of which are the errant wanderings of the human imagination.

Dreams and Nightmares: The Ambiguity of the Future

In many ways, what we imagine about the future is more revealing than analytical prognostications might ever be—not for what it tells us about the future, but about ourselves and our culture. Indeed, the future is the central subject matter of many art forms, including speculative fiction. This body of literature includes books that have become staples of the literary canon, such as George Orwell's *1984* (1945) and Aldous Huxley's *Brave New World* (1932). It also includes the books of popular science fiction writers such as Ray Bradbury, Isaac Asimov, Ursula Le Guin, and William Gibson. One hallmark of such work is its obsession with the technical gadgetry of the future, sometimes for its own sake but more often because of the impact technology might have on the future. This literary genre concerns itself with issues like mind control, consciousness altering, and social engineering. Looking at the genre as a whole—from H. G. Wells's *Time Machine* (1895) to Gibson's *Neuromancer* (1984)—we find a world of alternative visions, of brilliant light on one hand and impenetrable darkness on the other. We find not only naive adventure stories about explorers who cheerfully unlock the secrets of time and space, but also bleak tales of lonely individuals struggling against corporate hegemony in a landscape dominated by cyborgs. This dichotomy of attitude exists not just in literature but in futurist films and computer games as well. In the collective world of our cultural imagination, technology is used as often to oppress as it is to liberate the human spirit.

This same duality is also present in our daily lives. The journals of commerce are full of hope regarding the Age of Information. Advertisers tout the possibilities of the Internet. Schools turn their eyes on-line, heralding new curricula that will revolutionize education. These joyous imaginings, though, are not without fear. In news magazines and on television shows we also hear of technology's dark side. We hear that pollution from the manufacture of integrated circuitry is leaking into groundwater,[1] that more and more workers are suffering from carpal tunnel syndrome, that over the Internet information about our personal lives is made available to strangers or—worse yet—to the government. We learn that foreign enemies are scheming to use our satellite telecommunications systems to plot terrorist attacks. So in life, as in art, technology inspires the bright vision on the one hand and dark fear on the other. And this fear is reinforced by our knowledge that technology does not always work the way it is supposed to. Just like nuclear waste in an underground storage facility, the hazards of technology have a way of leaking out of their supposedly impermeable containers and infiltrating our lives.

THE DARK SIDE OF FUTURISM

This split vision of technology— savior one minute, demon the next—may have its roots in human consciousness, in an innate simultaneous pairing of fear and wonder of the unknown on the one hand, and an irresistible impulse to plunge ahead and explore regardless on the other. We can see this dichotomy in the myths and stories that transcend time and cultural boundaries, from the ancient Greek story of Pandora's Box to Mary Shelly's *Frankenstein* (1818).

This same dichotomy was also expressed in the work of the early twentieth-century Futurists, that group of writers and artists who developed the notions of collage and juxtaposition that have dominated the artistic sensibility of much of this century. This dichotomy is evident in a different way in the writings of Filippo Tommaso Marinetti, the Futurist writer who gave voice to an artistic fascination with technology and championed the ideas of multilinear lyricism and simultaneity as guiding structural precepts for artists. Even though it is important to recognize the contributions of these early Futurists, it is also important to recognize a darker, more complex side of the Futurist vision.

In 1909, when Marinetti wrote *The Futurist Manifesto*, Europe was moving toward World War I. Among the many causes of the war was the desire to break free of the feudal and monarchic systems of the past. As such, the Great War was seen by some as not only inevitable but necessary, as part of a schism with the past, a revolution of consciousness, a purging that was related to the revolution in artistic and social thinking engendered by the emerging new technologies. More than being necessary, war was seen not as something to be feared, but as something to be welcomed. "War is beautiful because it initiates the dreamt-of metalization of the human body. War is beautiful because it enriches a flowering meadow with the fiery orchids of machine guns," wrote Marinetti. "War is beautiful because it creates a new architecture, like that of the big tanks, the geometrical formation flights, the smoke spirals from burning villages."[2] Such sentiments were echoed by avant-garde artists throughout Europe at the time, by Guillaume Apollinaire and Blaise Cendrars and Umberto Boccioni. These artists' enthusiasm was tempered, however, by the deaths or maimings they witnessed in the war, or when they experienced the protracted, gruesome horrors of trench warfare.[3] After the war, the work of the Dadaists, Surrealists, and others working in the Futurist mode took a different tone. The techniques of juxtaposition and collage remain, but the machines frequently appear against landscapes of devastation that often include human forms defiled and transformed (see Figure 10.1).

Meanwhile, early Futurist rhetoric passed over into the mainstream. In the wake of World War I, the young Benito Mussolini—a powerful and charismatic orator in his own right—infused his own Italian nationalism with a sensibility and phraseology borrowed in part from Marinetti's fiery writings on Futurist art. Marinetti himself became a supporter of Mussolini, as did Ezra Pound and many others. Ideologically, Mussolini combined the Futurist passion for machines with corporativism, an economic philosophy that placed the capitalist titans of industry in close allegiance to a centralized government that was to maintain order and provide infrastructure for industrial overlords who managed the workforce and the economy. This philosophy, embraced by many businesspeople, including the American automaker Henry Ford, went by the name of fascism. Despite Mussolini's use of secret police and torture, fascism was regarded in a more or less positive light throughout the world until Mussolini's alliance with Adolf Hitler in

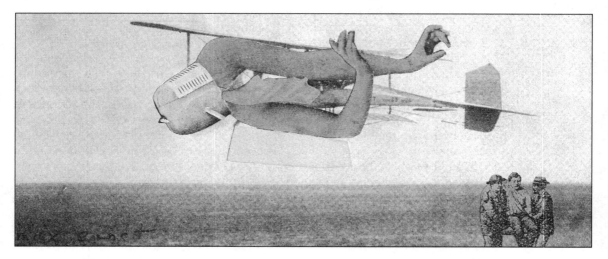

Figure 10.1
Max Ernst, untitled *(Deadly Plane),* 1920. This use of collage, while employing the artistic techniques inspired by the Futurist sensibility, expresses underlying uneasiness about the effects of machines on humanity. (© copyright 1997 Artists Rights Society (ARS), New York. Menil Collection, Houston, Texas)

the late 1930s, when it became associated with racism, concentration camps, and ethnic genocide. (Some writers on economics, such as Noam Chomsky and Michael Harrington, maintain that the central economic tenets of fascism—particularly in regard to corporativism—remain in effect today, even though some of the wording and semantics of the philosophy may have changed because of fascism's earlier association with genocide and totalitarian regimes.)

Thus we see that the Futurist vision that has dominated much of twentieth-century art and politics is by no means a simple one. It contains as much darkness as light, as much malevolence as hope. Just as Futurism has been an inspiration to some great artistic, social, and scientific achievements, it also has been linked to the creation of unspeakable violence. This same thing can be said for almost all political, religious, and spiritual movements. Hitler invoked Nietzsche to justify his vision of the super race, just as nineteenth-century Americans invoked Jesus Christ during the roundup and extermination of the Indians out west. The truth, perhaps, is that all ideas and philosophies are inherently based on contradiction, and that violence is a fundamental part of human life. With their fondness for radical juxtaposition and collagist techniques that emphasize paradox over linear rationalism, Futurist artists would be among the first to accept such contradictions as inevitable. Rather than fearing and denying this paradox, however, as do most adherents of orthodox religious and political systems, the early Futurists relished it.

So perhaps it should not be unexpected that technology and the future it will bring are often seen in contradictory terms. We can certainly see this same ambiguity in the polarized attitudes that dominate the current debate over interactive media.

DIRECTIONS FOR MULTIMEDIA

Its advocates see interactive media not only as the wave of the future, but as a step in the evolution of human consciousness. It is, they contend, a movement away from the hierarchical form of communications embodied by the teacher-centered classroom or by network-controlled news. It is instead a movement toward a user-centered experience in which users control the information. In this context, the advent of interactive media has been hailed as a democratization of the mass media because it takes the power of information out of the hands of corporations and special interests and gives it to the individual. In this utopian view of communications, everyone has equal access to information, the user is the speaker as well as the one spoken to, and the world is enriched by the free flow of ideas.[4]

Skeptics, however, see a much grimmer future. Rather than the democratization of ideas, they see a chaos of contending opinions that in the end will be dominated not by rational discourse but by politics and money. They see the extension of existing corporate empires into the realm of the computer, creating a world in which an Orwellian Big Brother in cyberspace tracks the comings and goings and the most intimate habits of its citizens. As more and more social functions move on-line, the computer becomes a tool for the ultimate centralization of information, capable of selling all our vital statistics—who we are, what we think, what foods we eat, what goods we purchase, and our political affiliations and reading habits as well—to the highest bidder. In this view, the home computer represents not freedom, but a tool for others to track and control our everyday lives.

The two extreme visions presented here, of course, are oversimplifications of positions in an ongoing debate that is actually even more complex. Even so, the current debate echoes the debates that have accompanied technical developments throughout the ages, from the harnessing of fire to the invention of the atomic bomb. At the heart of the debate, lies a simple concern: What will be the consequences of technological innovations?

We have seen that answering such a question is a risky exercise; it is not easy to predict the future. Human beings and the technologies they create tend not to behave exactly as expected. Unforeseen developments and unimaginable consequences often ensue.

The agricultural revolution in the mid-twentieth century, for example, trumpeted the wonders of chemical pesticides and fertilizers. These advances were supposed to end hunger in the Third World and facilitate the financial independence of developing nations. Neither of these expected benefits came to pass. The world today has more hungry people than ever. Furthermore, environmentalists have raised serious concerns about chemical agriculture and its long-range effects on the environment and human health.

Another example that is more closely related to the subject at hand concerns television. After World War II, newspaper and journal articles hailed the coming of television much as they now hail the arrival of new computer media. Many predictions were made: Television would bring higher education to the masses; people would attend educational classes in their homes: drama, music, and the arts would flourish thanks to new audiences in the living rooms of middle America. Now, fifty years later, few people have much good to say about television programming, particularly broadcast television. Instead of increasing community awareness, some argue, television has narrowed political expression to fifteen-second sound bites sanctioned by the networks and the corporate interests that control them. Educational television, some say, is a contradiction in terms.

This does not mean that advances in technology are without benefit; both modern agriculture and television have wrought profound changes, many of which have been beneficial. The point is that the nature of the changes caused by new technology is unpredictable. Even though many of these changes may be for the good, many of them are not. The same is likely to be true for advances in the new digital media.

Technical Advances in Interactive Media

If there is one article of faith among the computer cognoscenti, it is that technical advances will continue to expand the power and scope of the personal computer. If this belief proves justified, undoubtedly advances in compression schemes will soon allow for digital video to be delivered more easily into the home.

The main question, as we have seen, is what the avenue of delivery will be. Will it come through the Internet, with fiber-optic cable carrying full-motion multimedia to personal computers, or will the end medium be the television set, with interactive programming delivered via cable into the home entertainment centers of America? Or will some other combination of delivery media be involved?

Such questions are not trivial because behind the scenes lies a struggle among public utilities, broadcast companies, and computer corporations that has as much to do with political maneuvering and hardball economics as it does with technology. The results of this struggle will greatly affect the future of telecommunications. It will determine not only how information comes into our homes but who will deliver it and what it will be.

Computer screens and television screens, though physically very similar, are approached by different audiences with different psychological expectations. Televisions are a source of entertainment; computers are associated with work. Televisions in America are located at the hearth; people relax while they watch TV, they recline, they share it with family, they eat. Computers, on the other hand, are located in offices and dens; people lean forward into the screen, usually alone, typing, entering data, manipulating information. The two environments have different ergonomics, different social functions, and different participants.

If the primary delivery medium for interactive programming ends up being television, what does this imply? More interactive shopping and less educational material? If, on the other hand, the computer wins out, what does this portend?

In the end, there may be one delivery medium but two platforms. In other words, audiences may have access to interactive communications in both the family room and the office, but the machines would have different interfaces that enable the performance of different tasks. As computers, telephones, and televisions come to resemble one another and eventually merge into one device, we begin to wonder how these devices will be controlled, and who will control them.

Everything costs something; in this case, what will the price be? Will the audience pay for access, or will the price be interactive advertising delivered in endless obligatory cuts? Will part of the price be submission to government watchdogs and corporate demographers who monitor audience behavior for their own political and economic gain? And what are the social costs of moving toward a virtual communication system that in a very real sense values image above substance?

Ours is not a society that reflects on such issues for long. Rather, we are more likely to embrace change first and worry later. To see that this is true we need only consider the effects of the automobile, which changed not just our transportation infrastructure but our physical and social landscape as well.

Even though we are aware of the more obvious negative side effects of the automobile—pollution, congested roadways, urban sprawl—our tendency is to view these outcomes as inevitable. But the results stem not from inevitability, but from the way our society operates. Until the early 1950s, the large cities in the American West—Los Angeles, the San Francisco Bay Area, Seattle, and Portland—had well-developed public transportation systems that included electric buses, trolley lines, and trains. Then General Motors and the other automakers bought these systems and destroyed them, tearing up the tracks and lobbying for new freeway construction. The reason, of course, was not because anything was wrong with the existing systems. The reason, instead, was to eliminate the automotive industry's competition. With public transportation no longer available, consumers would buy more cars (and more tires, and more gasoline and oil). Such action may have generated more profits for some business interests, but it also destroyed a valuable public resource that has proven very difficult to replace.

Some might argue that the corporations of today no longer behave in such a ruthless manner, that they have a heightened sense of altruism and greater social consciousness, and that in any event our government is watching out for us. History would suggest, however, that these are naive assertions. The shape of interactive communications is less likely to be determined by altruism and concern over public good than it is by the political and economic interests of media corporations, large computer companies, and public utilities vying to place themselves in the most profitable position possible.

The Multimedia Marketplace in the Future

We have already noted how the enthusiastic proponents of the Digital Revolution view interactive media as a potential means for transforming communications. They envision a shift to a user-centered paradigm in which the audience not only gets wide access to information, but also can participate in the dissemination process by creating projects that can compete with much larger and more powerful, established commercial entities. A potential obstacle to this vision looms in the brute forces of market economics, which places value not on the individual but on consumption by the masses.

As mentioned earlier in this text, the notion of user control is somewhat misleading. In most interactive media, users do not get *control* of the material so much as they get *selective access*. The nature of this access, and what kinds of selections are available, are usually controlled by the program. Thus it is the designer who sets up the paradigm that determines the user's experience. Just as it is possible to control content through obligatory cuts and interactive structure, it is also possible to control access by demanding passwords, the payment of money, or other forms of behavior.

Enthusiasts also argue that the Digital Revolution places distribution and production into the hands of users, making it possible for people to create their own media and distribute them in ways previously reserved for large media entities. There may be some truth to this, but there are also two related problems: production quality and distribution. Mass audiences have grown use to high-end production quality, particularly in media projects. Such quality requires both sophisticated skills and the money to pay for those skills. CD-ROM projects, for example, can be expensive to produce and often require a wide range of personnel, from programmers to writers to visual artists to cinematographers to actors. It is rare for a home-based operator to have all the skills and resources necessary to attain high-end production quality. Still, sometimes high-end production skills are not necessary or the content itself is so dynamic and powerful that it supersedes matters of technical presentation. In such instances, producers still face distribution problems. Even though digital media may provide increased access to production, so far it has done little to expand distribution channels.

If anything, the conventional marketplace has experienced a narrowing of distribution channels over the past half century. The mass market is dominated by chain stores that rely on distribution systems that emphasize the mass sales of single items. This is as true for retail clothing stores as it is for bookstores, grocery stores, and office supply houses. The situation among independent stores offers some opportunities, but these businesses, too, rely on the same supply lines. For interactive media, the distribution system has greatly affected the nature of the CD-ROM market, in which the number of titles far exceeds shelf space and the current emphasis is on large studio productions that reintroduce characters and material with which the audience is already familiar. Rather than opening up the market, interactive media is being distributed along established channels that in many ways are even narrower than those used for books and video.

In the end, the Internet may offer the best chance to undermine the classic distribution paradigm. By posting products and services directly on the World Wide Web, individuals with low budgets theoretically can participate in the same forum as large corporate entities. In practice, though, this has proved difficult. Although it is possible to produce elaborate and beautiful Web sites, notifying the audience of their existence is another matter. Most large Web sites rely on traditional media—television, newspapers, and magazines—to notify users of their existence. Meanwhile, the large software companies are working on paradigms to centralize the Web-browsing experience. Their purpose is not to open the distribution system to independents but rather to build a commercially viable interactive structure in which they can control the audience's movement, sell advertising, and charge access fees to users.

Despite a good deal of political rhetoric to the contrary, success in the marketplace is not the only indicator of social worth. As we all know, many financially successful products have very little social value at all, and some, like cigarettes, are even quite harmful. The market may be a good indicator of popular sensibility —or, more accurately, a good indicator of the ability of advertisers to manipulate popular sensibilities. Even so, the market is not necessarily supportive of other worthwhile attributes, particularly if these attributes happen to offer a lower profit margin or are in some way contrary to the dominant sensibility.

Thus in regard to market access, technology closes down opportunities even as it opens them up. More people will have access to the tools to create sophisticated media presentations, but finding an audience in an environment as diffuse as the Internet is no easy matter. Even so, this diffuseness, this lack of centralization, may be just the climate in which certain types of individual expression may flourish outside the corporate mainstream. In either regard, whatever possibilities and hindrances lie ahead, one of the challenges facing the creators of new media is producing intelligent, worthwhile programs in a market whose values and interests often lie elsewhere.[5]

The Future of Interactive Education

Despite the limitations just discussed, it is difficult not to get excited about the future possibilities of interactive education. Part of the excitement rests in the hope that schools will be able to reach more students using fewer resources. Online technologies, educators hope, will help reach out to new audiences, perhaps even bringing new opportunities to the socially disenfranchised.

In Chapter 7 we discussed some of the limitations of computer-assisted education, including initial development expense and the need for constant updating of software and hardware. To date, these requirements, as well as the need for technical expertise, have tended to narrow the audience rather than expand it. Before technology can be successful in an educational context, such fundamental concerns need to be addressed. Equipment must be standardized before access can be universal. Unfortunately, the computer companies have little incentive to achieve any true standardization, for a great deal of their commercial energy is devoted to selling extensions, add-ons, upgrades, and updates, of software and hardware alike, most of which lack compatibility and require a great deal of time, patience, and technical expertise on the behalf of users.

Regardless, the hope remains that interactive multimedia can help institutions reach larger audiences more effectively. Just as universities turned to videotaped lectures to teach required courses in the early 1970s, a similar movement is now afoot. If this movement continues to develop, it is likely that multimedia and on-line courses may soon supplant traditional instruction, at least for some subjects in some institutions. Even so, there will be limits. In many large state universities today, undergraduates have little direct interaction with teachers during their first two years. Instead they are taught in large lecture halls, often by videotape, and their exams are graded by teaching assistants. If multimedia and on-line curricula can add dimension to this situation by freeing up teacher time, it might be successful. If, on the other hand, it results in the further depersonalization of higher education, the long-range results could be disastrous.

Similar hopes and fears exist regarding the use of computers in elementary and secondary education. One hope is that computer-aided education will help alleviate the financial strain on public schools as they struggle with declining budgets and anticipated increases in enrollment. Again, the implementation of interactive educational programs may help students work their own way through required material, freeing teachers from repetitive work and making them more available for direct interaction with students. Legitimate concerns, however, exist in the fears that the financial costs of technical implementation may be greater than anyone recognizes; that faculties will become smaller and their duties more administrative than educational; that the overall quality of education will suffer as the emphasis turns toward the production and use of media and their technical upkeep, rather than on one-on-one interaction with students.

Proponents of computer-assisted education argue that computers will allow for better access to information and education for all, regardless of class and social boundaries. Opponents fear that the computer will not be a democratizing tool but precisely the opposite, that our society is increasingly becoming a society with only two classes, the haves and the have-nots. In such a world, the poor will be shackled not just by their general lack of literacy, but by their lack of technical literacy as well.[6]

If this is true in the United States, it is even truer abroad, where the majority of the world's nations, particularly in the Third World, have suffered decreases in standard of living over the past quarter century. If this trend continues, advances in technology may only serve to widen that gap as emerging nations find themselves increasingly unable to compete.

Computer and telecommunications companies have very carefully crafted public images. Their advertisements emphasize that their products break down barriers and bring the world's peoples closer together. Such claims have their elements of truth, particularly if one considers business communications among the affluent nations. In a wider sense, though, there is little to suggest that large computer and telecommunications companies have an interest in removing social barriers, or even in acknowledging their existence. Despite its liberal veneer and global presence, the computer industry is more renowned for its self-absorption than its social largesse. The industry as a whole is generally indifferent to the notion that it should take responsibility for changes caused by its persistent expansion. Instead, there is a general belief within the industry that it is improving the world simply by creating new devices. Given such attitudes, there is a great risk that the suffering and exclusion of the great proportion of the world's people will not only go unaddressed but unrecognized, as they have neither the resources nor the technical skills to make their voices heard on the new worldwide medium.

The Direction of Society in an Interactive World

In Chapter 1 we discussed how the activities of a small group of artists foreshadowed the perceptual changes that have dominated the intellectual life of the twentieth century. As we have seen, the early Futurists foresaw the cross-disciplinary fusion of the arts and sciences in the service of human expression, and it is in their writings that we first find expressions of the concepts of multilinear lyricism and simultaneity. The Futurist movement was influential not just in the arts, but in literature, science, and the humanities. It was concurrent with a philosophic movement away from Newtonian and deistic views of the universe, replacing them with a cultural and scientific relativism.

The current fascination with multimedia is an extension of this century-long movement. Even the interdisciplinary production process for multimedia—the fusion of techniques borrowed from cinema, music, literature, art, computer games and software design—is itself a fulfillment of the Futurist vision. More importantly, perhaps, is how thoroughly multimedia enthusiasts embrace the relativistic view of the universe in which randomness, individual exploration, variable point of view, and unexpected juxtaposition are inherent to the experience of the medium.

Still, the question remains: Where will this new medium lead?

Speculating on technical possibilities is in many ways easier than anticipating social consequences. Given the events of the past decade, it does not seem risky to suggest that compression and transmission capabilities will continue to improve. Also, with the recent development of flat-screen television, high-definition monitors, surround sound, and voice recognition, the technology for creating more intense, more fully immersive multimedia already exists. As such capabilities increase yet further, the trend will be toward the creation of virtual environments that blur the distinction between fantasy and reality. This will no doubt spawn a new generation of simulation devices, some of which will be used for training and education, others more for entertainment.

In the future, the interface between human and computer is likely to become more intimate and transparent, taking advantage of the computer's increasing ability to respond to individual voices and gestures. Electrode implantation may eventually allow for a direct link between the human brain and the computer, making the computer a true extension of the mind, a repository for ideas and documents, and for fantasies as well.

As always seems to be the case, our ability to create technical devices runs ahead of our ability to use such devices wisely. In the case of the computer, perhaps the most serious question concerns its potential for blurring the lines between artificial and human intelligence.

It is part of human nature to alter reality, just as it is natural for the rational mind to fear change. The struggle between fantasy and reality, between the natural and the artificial, is part of the creative act and is inherent to the arts. Because part of the job of the artist is to envision alternate realities, artists often elicit criticism, particularly if their visions threaten the perceived reality of the dominant society. Sometimes, perhaps, people are right to feel threatened, because the power of art is very great. Even so, social change is not the primary purpose of artistic creation. Rather, art provides a kind of wormhole into the fantastic, a kind of psychic journey in which humans explore their hidden nature and then return, enriched, deepened, somehow more in touch with the mysterious forces of life.

Despite their obvious differences, books and movies offer similar experiences, submerging their respective audiences into the richness of other worlds. With the cultivation of virtual technologies, computer-generated media will offer experiences every bit as rich and intimate. The challenges for the writers and designers of programs for the emerging media will be the same challenges artists have faced for centuries. They will have to struggle with questions of meaning and message, with matters of spiritual and moral concern, and with understanding—for themselves as much as for their audience—of the true nature and value of the work they are creating.

Perhaps the dawn of interactive media is symbolic of a dawning new human consciousness that is global, democratic, and user-centered, that breaks down the old patterns of oppression and lets loose new patterns of thought. This view is certainly widely espoused in commercial magazines and programs of various stripes, all aimed at the so-called digerati, the new generation of intelligentsia whose countercultural hipness and technological savvy are juxtaposed in apparent anticipation of a new age that places the computer at the center of our lives.

Still, it is important to be skeptical—to look at who is in control behind the scenes and who is paying for the advertisements; to recognize that the veneer of hipness is often nothing more than hype, a means of selling the computer companies and their products to an ambitious, affluent audience; to realize that underneath the technoglitz we may be building a society as decadent and culturally bankrupt as that portrayed in William Gibson's *Neuromancer*.

Undoubtedly wondrous interactive programs will be created. New training simulations will allow doctors to save more lives, architects to design better buildings, scientists and thinkers to explore new realms. Equally true, no doubt, is the fact that many people will not be reached. Others will be bypassed by the new technology or become victims of the psychological and social pollution strewn in its wake. Just as there will be success in interactive education, there will be outlandish stupidities. Just as great works of art will bloom, moronic and idiotic ones will be spawned as well.

There is no reason to expect that the world will either grow any saner as the result of technological implementation or become inherently any worse. This does not mean, however, that we should not try to make sense out of what lies ahead, or that we should abandon our responsibilities. To say that one goal of the Digital Revolution is to promise a multitudinous view of reality does not mean that we can avoid those views of reality that are unpleasant to us.

The paradox of relativism lies in its assertion that every point of view can be considered as legitimate as any other. Although this may allow for the emergence of new sensibilities, it also creates problems. If all things are equal, then what is true and what is false; what is good and what is evil? What is the best way to proceed when all paths have the same value? The idea that each human being contains not a single self, but many selves, may be liberating at first, but human consciousness seeks always to define itself—to sort, to arrange, to sift out meaning. The establishment of meaning means making choices, choosing directions, and establishing hierarchies, and the existence of a hierarchy implies the end of a relativistic view. This is the great paradox of relativism—that it constantly gives birth to its own demise. A relativistic view cannot hold, but neither can a hierarchical one because our experiences tell us it will always be undermined.

In the face of this paradox, there are two possible reactions. One is cynicism—to declare that because there is no truth, any truth will do, especially if that truth is convenient in the moment and allows us to get the things we want for ourselves. The other, more difficult course is not to be beaten down by the paradox, but to be informed by it. Truth is not constructed but rather sought, and the seeking is continual. Truth and falsehood exist simultaneously in any assertion, but this does not mean they are one in the same thing. There are subtleties, there are gradations, there are angles of perspective. Capturing these subtleties and conveying them in context is part of the writer's job, no matter the medium. The courage not to self-delude; to present things from many angles, to include the left behind and the forgotten as well as those with the world at their fingertips; to extract from the confusion a unifying message whose intention is not to oppress but to illuminate—these are the essential goals of writers not just in multimedia but in all media, no matter the technology.

SUMMARY

Interactive media are, in part, outgrowths of technical advances in telecommunications and the computer sciences. Like many changes shaped by technical advances, these new media are double-edged swords. They offer hope for the future, yes, including improved ways of communicating, opportunities to break down barriers, and a means of spreading knowledge and expanding human consciousness. At the same, though, these new media threaten to divide society yet further: to bestow blessings on a privileged few, to dehumanize, to obfuscate by focusing on form as opposed to content, and to separate people increasingly from the physical world and the consequences of their actions.

These conflicting views of new media, diametrically opposed as they may be, have antecedents in both the history of ideas and the empirical world of observed phenomena. There is plenty of evidence to suggest that the blessings of technology are mixed. The philosophical outgrowths of early Futurist thought, for example, were intertwined not only with some of the most exhilarating and liberating perceptual changes of the twentieth century, but also with some of the most frightening social movements, including fascism and a Neitzschean obsession with totalitarian regimes. Similarly, the social consequences of industrial technology have been equally ambiguous, improving daily life for many but at a considerable cost, including widespread social alienation and ongoing degradation of the environment. Similarly, twentieth-century advances in communications, such as television, have not always lived up to the unabashed hopes of their early promoters. Even though the television has brought the world closer together, its message has not always been an intelligent one. In fact, many condemn it as a device that discourages intelligent thought and instead spreads an increasingly banal and monolithic message to an ever widening audience.

Like television, interactive media offer similar promises to revolutionize communications and improve society. They must overcome some serious challenges, however, if the promised changes are to be for the better. Among those challenges are the current structure of the marketplace, which has seen a narrowing of distribution channels over the past few decades. Because this distribution system tends to view all human behavior (including the act of communication) as a commodity, there is a tendency to value programming according to potential profits rather than transcendent values having to do with social, cultural, or spiritual worth. Of further difficulty is the nature of the computer industry itself, which traditionally has been obsessed with the technology for its own sake at the expense of any social consequences. Despite these difficulties, though, the technology offers hope to educators seeking to transcend the walls of the traditional classroom. Though large commercial forces are fighting for the control of the Internet, its inherently diffuse nature may ultimately foil those efforts. Perhaps the Web may live up to its promise and engender forms of communication and distribution that are less dependent on traditional means.

In the end, though, the new media are likely to fulfill the prophecies of their supporters and detractors alike. Media are human instruments, after all, and like all human instruments they can uplift or they can degrade and defile humanity. In this regard, it is up to the people working in the media to be aware of both the dangers and the opportunities that lie ahead. Even though it is true that the political and economic forces of the mass market are often far stronger than individuals, it is also true that on a daily basis it will be individual writers and designers who shape the content that will drive the new media. In other words, it is up to these individuals to take responsibility for the various and multitudinous worlds they seek to create.

EXERCISES

1. This chapter poses two alternative views on the long-range effects of interactive media, one positive and the other negative. Which do you see as being the more likely? Give examples to support your contention.

2. Describe some of the hidden social costs of interactive media. What kinds of interactions and ways of doing things might we lose? What valuable things should we preserve?

3. In this chapter, the author describes how General Motors deliberately dismantled public transportation systems in western U.S. cities in order to spur automobile sales. Do you see any analogous situations developing in regard to interactive media? (In other words, have you noticed any useful systems or beneficial ways of doing things that are now being phased out in the name of the information superhighway?) What are the motivations of the people advocating the changes? Of those resisting?

4. What advantages does interactive education offer over traditional education? What disadvantages? Are there ways of using both new and traditional methodologies together? If so, describe them.

5. What one thing makes you most uneasy about the presence of computers in your life? What might you do about it?

6. Define the concept of relativism. Describe what this notion has to do with the writing and design of interactive media.

7. What do you think the writer's role in a multimedia project should be?

ENDNOTES

1. Jane Kay, "Bay Area's Worst Pollution: Mess from the Past Continues to Plague Silicon Valley Water," *San Francisco Chronicle*, December 1, 1996, A12. According to this article, chemical by-products of the microchip manufacturing process, such as trichloroethylene and Freon 13, now contaminate the aquifer in much of Silicon Valley. Near the IBM plant in South San Jose, for example, underground contamination stretches 3 miles and "has forced the closing of 8 private wells and 17 public wells, which had once served 100,000 San Jose residents." The irony is that such industry was originally welcomed as nonpolluting, "clean, light" industry, good for the economy, and safe for the environment.

2. Marjorie Perloff, *The Futurist Moment* (Chicago: The University of Chicago Press, 1986), 30.

3. *Ibid.* See also Sidra Stitch, *Anxious Visions: Surrealist Art* (New York: Abbeville Press, 1990).

4. John Brockman, *Digerati: Encounters with the Cyber Elite* (San Francisco: HardWired, 1996). See also Kevin Kelly, "What Would McLuhan Say?" *Wired*, October 1996, 149.

5. "The Web Maestro: An Interview with Tim Berners-Lee." *Technology Review*, July 1996, 38.

6. *Ibid.*, 40.

Script Excerpts from the Interactive Cinema *In the 1st Degree*

In the 1st Degree is an interactive courtroom drama in which the player assumes the role of a prosecuting attorney, following the case from the murder scene, through the pretrial interviews, and finally to the trial itself. The goal of the player is to get the jury to return a guilty verdict.

Following are excerpts from several phases of the scripting process, including the story overview and character bible, both of which were part of the original treatment. Also included are dialogue excerpts from the production script and a sample flowchart used to create dialogue exchanges between the player and the characters within the program.

This program was produced for Broderbund by Adair & Armstrong, an interactive design group with a background in documentary filmmaking. The lead writer was Domenic Stansberry.

Excerpt from the TREATMENT: Story Overview

Not long ago in San Francisco—on a morning that was blustery and pale, when the fog showed no signs of lifting—popular gallery owner Zachary Barnes was shot dead in his art gallery. Barnes's longtime friend and business partner, the artist James Tobin, was taken into custody at the scene, and—after a grueling interrogation—charged with murder in the first degree.

As the trial approaches, the office of the prosecuting attorney and the police have both gone silent, as is their custom in such cases. The official silence, of course, only encourages unofficial speculation: Was this art theft, a business deal gone bad, a love triangle? Everyone in the city, it seems, has their own explanation for what happened, and with each passing day these stories become more involved and entangled, so that the scandal—in the minds of some—permeates all levels of society, up into the mayor's office, down into the street, becoming a palpable entity in the life of the city, all the more so because the exact truth is unknown.

The job of the prosecuting attorney, Sterling Granger, is to reduce these many stories to one story, or more precisely a single phrase that can be written upon a piece of paper and read aloud by the foreman of the jury: "We, the jury, find the defendant . . . *guilty.*"

But this will not be easy. Truth in a jury trial, as in real life, seldom emerges so cleanly; rather, it is a series of vignettes, of possibilities, of half-glimpsed scenes that the attorney must weave together into a picture of events so clear that it can elicit that single phrase.

In the trial, the player is Sterling Granger, and as such the player experiences the story from a multiplicity of angles while reviewing videotape and interviewing witnesses in a Rashomonesque process that explores the tentative nature of truth, innocence, and guilt. In the role of Granger, the player searches for the story that will lead to conviction, knowing all the while that the other participants in this drama—the defense council, the key witnesses, and especially the defendant himself—are pursuing their own agendas, choosing from a vast array of stories their own particular story, and preparing to tell it before the court.

Excerpt from the CHARACTER BIBLE

There are several key witnesses in this case, each of whom are video-taped by Inspector Anita Looper immediately after the murder. Acting as Granger, the player has the opportunity to view these tapes, examine the witnesses before the trial, then to call them to the stand during the trial itself. All of these witnesses are well-developed characters, with their own interior conflicts and dilemmas and their own reasons for seeing the story their own way. It is the task of the player to induce each witness to offer testimony favorable to the prosecution.

CHARACTER NO. 1: RUBY GONZALEZ

Background: Ruby is a twenty-three-year-old art student who met Tobin when he was a guest lecturer at the Art Institute, where Ruby is attending classes on scholarship. She is from a workingclass neighborhood of San Jose. Though she has a considerable amount of natural talent and a kind of street bravado, she lacks self-confidence, especially in her ability to maneuver in the art world. She is attracted to the glitter but is also out of her element, and her attraction to Tobin is partly an attraction to the ease with which he moves through this other world. In the beginning, he takes an interest in her art work as well; he is a good looking man who has an ironic sense of humor and a dedication to his art. As time goes on, however, Ruby comes to feel that he is using her: that his real devotion is to himself, and that even his desire for artistic success is not so much a passion for art as it is a serving of the self. She begins to realize that for Tobin—sex, love, friendship, and art—all these are an elaborate aggrandizement of the self. There are even stories among Ruby's friends that he is abusive to her, though Ruby herself backs away from these accusations.

In this emotional milieu, she befriends Tobin's business partner, Zachary Barnes, a thirty-five-year-old black gallery owner who, at a younger age, was not unlike Ruby: an outsider trying to penetrate a forbidden world. He too takes an interest in Ruby's work. She confides in him about her problems with Tobin, and Zack—who is having some problems of his own—confides in her. They end up sleeping together, an event that triggers a confrontation between the two men and appears to be at least a partial motive in the murder.

Ruby's Dilemma: After the murder, Ruby is faced with a decision: Does she offer evidence against her boyfriend, James Tobin? Or, to phrase it another way, does she choose Tobin or Zack? In some ways this exterior dilemma mirrors Ruby's interior dilemma in that Tobin represents a manipulative presence she must overcome in order to move toward self-realization.

The Prosecutor's Task: At the onset of the investigation, Ruby is reluctant to offer any information that is harmful to James Tobin. It is the task of the prosecutor to overcome that reluctance, and this is done largely by playing "good cop": by flattering her, yes, but more importantly by nudging her toward those things that seem to engender a positive vision of herself. If handled appropriately, she can offer key evidence concerning the murder weapon, and the fact that she had seen the gun in Tobin's possession prior to the murder. She can also undermine Tobin's alibi in regard to the painting theft and provide testimony suggesting he had planned not only the theft but Zack's murder as well.

CHARACTER NO. 2: SIMON LEE

Background: Simon Lee, also in his early twenties, is an assistant at the gallery who came to work for Tobin and Zack about eight months before the murder. Originally from Florida, he dropped out of school and came to San Francisco, where he has been living in the lower Haight. He is a tough kid but vulnerable, wears tattoos, and hangs the mean scene on the streets.

Before coming to work for Tobin, Simon worked for a messenger service, making deliveries on a motorbike, dashing up and down the streets, but he walked away from his job one day and into Tobin's gallery. Though he maintains a tough exterior, he actually feels indebted to Tobin for taking him and admires the man's passion and success. Tobin also has been letting him work as a painter's assistant, stretching canvas and doing preliminary background painting on some of his work.

Simon had a key to the gallery, and late one night while out carousing with some friends he comes to grab a jacket he's left behind. There's a racket in the main showroom, and Simon discovers Tobin stealing his own paintings. Tobin sees Simon as well and tells him to keep his mouth quiet. Later, some extra cash appears in Simon's pay envelope.

Simon went along with this—it was just insurance money anyway, he figured—and it was none of his business. But when Zack turned up dead, he began to have second thoughts and questioned his loyalty to Tobin.

Simon's Dilemma: Like Ruby, Simon was drawn to Tobin by the strength of Tobin's personality. After the murder, however, he realizes he has been a poor judge of character to have allowed himself to become too involved with Tobin's schemes. He would like to come clean, but to implicate Tobin he must also admit his own complicity in events—silently taking the money (as well as inadvertently helping Tobin secure the murder weapon). This is not an admission he wants to make.

The Prosecutor's Task: To succeed with Simon, the prosecutor must help the young man overcome his fear of implicating himself. Unlike Ruby, Simon responds more to threats than to cajoling; he has a fear of going to jail. The prosecutor needs to push on Simon, because if he goes easy on him, Simon will try to skate away from the truth. Simon can help the prosecution concerning the theft, obviously, and he can also help tie the murder weapon to Tobin.

Excerpt from the PRODUCTION SCRIPT

Opening Sequence

INSPECTOR LOOPER INTRODUCES THE CASE

We are in a cheap diner; Anita Looper is sitting at the counter eating a large breakfast. As we sit down next to her, she turns to us (addressing the camera directly) and begins talking. . . .

LOOPER: How'd you know where I was?
GRANGER: Like you're not here every morning?
LOOPER: Long as nobody's murdered nobody. Starting on the Barnes case today?
GRANGER: Yup.
LOOPER: I've gone over my videos—you know, the interviews with the witnesses—and snatched out the juicy parts. Copies on your desk.
GRANGER: Thanks.
LOOPER: An' one of them was the defendant. Good thing I talked to him 'cuz Charleston, his lawyer, sure's not going to let him talk to us now.
GRANGER: You mean Cynthia Charleston's the defense?
LOOPER: (sarcastically) Yup! More rubles than scruples, that one. You know, City Hall's already started calling over here.
GRANGER: How come?
LOOPER: (shouting) Sal, what happened to my side of bacon already? Victim's widow works for the mayor. Press officer.
GRANGER: (sarcastic) Great!
LOOPER: Wait, more good news. I'd pretty much figured this an easy case: two guys, business partners, one of them schtups the other's girlfriend, he gets just a little mad and . . .
GRANGER: (interrupting) Wait, how do you know this?
LOOPER: Check your "documents file," you'll see. Anyway, seems pretty clear the two guys got into a fight and only one walked away. Your job's gonna be to prove he planned it that way all along. Not that I want to tell you how to do your job.
GRANGER: Course not. What do you know about the grand theft larceny charge?

LOOPER: Not much, but more than anybody else. Seems a whole bunch of paintings were "stolen" from the gallery a little while before the murder. Tobin and Barnes were going to split the insurance money. Close to half a million dollars.
GRANGER: So maybe this whole thing started with a fight over that theft? Maybe the deceased got cold feet?
LOOPER: He does now. Pass the syrup, please.

Excerpt from the DIALOGUE SCRIPT

Game Phase: Pretrial Investigation

Character: Simon Lee

Topic: Gun

We have tracked Simon Lee down to his new workplace, a courier service housed in noisy warehouse in the South of Market district of San Francisco. He is not happy to see us and is reluctant to talk about the case. He is wearing a T-shirt with razor-cut sleeves; his arms are tattooed, his nose and ears are pierced. We have chosen to begin the conversation by asking Simon about the gun. At any point in the following dialogue (charted in Figure A.1), the player may be break off this conversation, switch topics, or go to another witness.

SQ0001
Gun in gallery?
GRANGER: Did you ever see a gun in the Barnes Gallery?

SA0001
SIMON: (insolent) Listen, I just worked there. I don't go nosing around.

SA0002
SIMON: I don't remember.

SQ0002
Put him at ease
GRANGER: (patiently) Simon, you're not a suspect in this case, so there's no need to worry. As far as I know, you were just an employee at the gallery. Now, can we continue?

SA0003
SIMON: No problem. (beat) But like you said, I just worked there. I don't know everything that went on.

SQ0003, SQ0005
Press hard
GRANGER: Simon, this is a murder case, and I'm not in the mood to play games. Did you ever see a gun in the gallery?

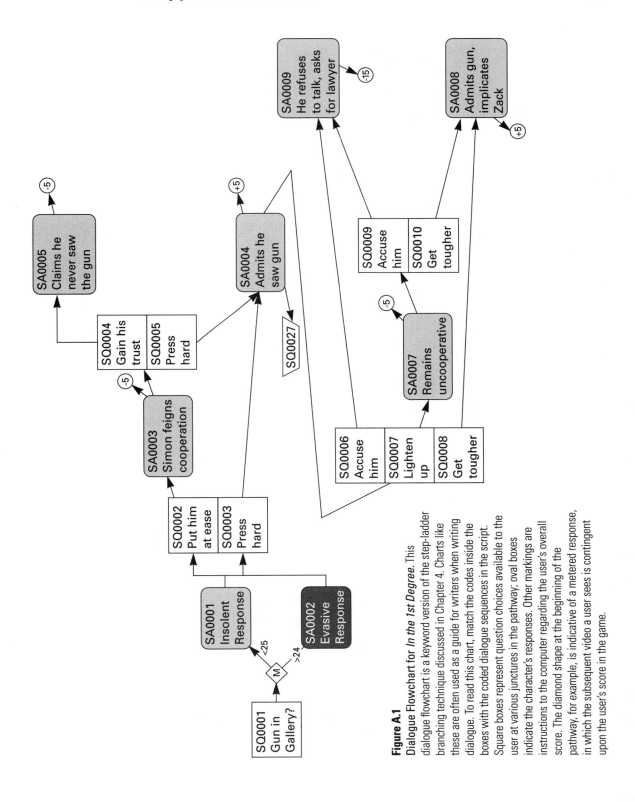

Figure A.1

Dialogue Flowchart for *In the 1st Degree.* This dialogue flowchart is a keyword version of the step-ladder branching technique discussed in Chapter 4. Charts like these are often used as a guide for writers when writing dialogue. To read this chart, match the codes inside the boxes with the coded dialogue sequences in the script. Square boxes represent question choices available to the user at various junctures in the pathway; oval boxes indicate the character's responses. Other markings are instructions to the computer regarding the user's overall score. The diamond shape at the beginning of the pathway, for example, is indicative of a metered response, in which the subsequent video a user sees is contingent upon the user's score in the game.

SA0004
SIMON: I'm not playing games.
GRANGER: Get real.
SIMON: (long beat) In Zack's desk—the cash drawer. It's usually locked. But about a week before the murder, he started keeping a gun in there. A short barrel. A .38, I think.
GRANGER: You didn't mention this to Inspector Looper?
SIMON: She talked to me right after the murder. I was shook up, and confused.

SQ0004
Gain his trust
GRANGER: I'm grateful for your cooperation. Are you sure you never heard either man refer to a weapon?

SA0005
SIMON: Nope. I never heard nothing about a gun. (He shakes his head, apologetically, as if he would like to help.)
GRANGER: So you never saw a gun, in the gallery? (beat) And you never heard anyone mention one?
SIMON: I'm sorry. No.
GRANGER: I'm sorry, too.

SQ0006, SQ0009
Accuse him
GRANGER: (sarcastic) I just had a wild thought, Simon. Maybe you brought the gun to Tobin?

SA0009
SIMON: You're nuts.
GRANGER: What did Tobin promise you, Simon? (mocking) A piece of the gallery?
SIMON: No. Two pieces. (insulted) How stupid you think I am?
GRANGER: I don't think you're stupid, Simon. I think you're a sap. It's written all over your face. (long beat while we study Simon's face)
SIMON: (lets out a breath) Maybe. But this sap's not talking anymore. Not about the gun anyway. Not without a lawyer.

SQ0007
Lighten up
GRANGER: The murder shook up a lot of people. It's understandable you would be confused. But about the gun—is there anything else you'd like to share with us?

SA0007
SIMON: No. (He smiles, with a slight smirk, as if mocking us.) I don't have anything else to share.

SQ0008, SQ0010
Get tougher
GRANGER: So you lied to Looper? Don't do it again. Now tell me, did you ever actually see Zachary with this gun in his possession?

SA0008
SIMON: Yeah. I saw Zack with it.
GRANGER: When?
SIMON: A few days before the murder. Tuesday morning? (shrugs, as if unsure) I got to work late and saw Zack standing at his desk. The guy unsnaps his brief case—and slides a gun into the drawer. (He makes a rapid motion, pointing his finger as if it were a gun, twisting it, sliding it into the drawer.)
GRANGER: What'd you do next?
SIMON: I just turned and left. I don't think he even saw me.

When done with this path, the player can choose to interview Simon on other available topics or proceed to another witness.

B

Excerpts from the Production Script of the Children's Educational Program
The Little Red Hen

The Little Red Hen is an interactive multimedia program for school children. It was produced by Computer Curriculum Corporation (CCC), a provider of K–12 educational software for use in classrooms and labs in elementary schools, in either networked or stand-alone computer environments. This particular program uses the traditional storybook as its central design metaphor, and it contains a broad base of visual instructional activities, open-ended opportunities for personal expression, and sound embellishments designed to enhance educational values.

The multimedia storybook gives young children options based on their own reading abilities. They can move quickly through the story, listening to narration and sound effects, or they can read the story interactively, clicking on buttons and hot spots for additional instruction. As they read, children can also record their own reactions to the story, using either an on-line microphone or an on-line notebook. After reading, children complete a number of creative activities to check their comprehension and respond personally to what they have read.

Included here are excerpts from the five major parts of the production script. Part I, from the opening sequence, is the script for the introductory screen, which presents the title of the story and introduces the main characters. Part II presents material from the pre-reading, or background activities; this includes a vocabulary exercise. Part III, the interactive read, presents script material from two scenes of the interactive story itself; each scene is analogous to a page in a storybook, and students are able to read and listen to the text pictured on the screen. Part IV includes some comprehension exercises, in which students answer questions about the story. Part V, from the closing activities, gives students a chance to register their feelings about the story.

In reading a script of this nature, it is important to understand certain scripting conventions. In educational programs, it is common for each major section to begin with a statement of objectives. Part II, for example, from the vocabulary section of the pre-reading activities, includes its educational objective at the outset. Following this, the subject matter of each particular screen is revealed in key words and phrases that will be the educational focus of that screen. (The educational focus of Vocabulary Screen #2, for example, is action verbs.) The statements regarding objectives and focus are for the benefit of the design team; they will not, of course, appear on-screen during the program itself.

The following information about the layout of the script is also useful:

SCRIPT AND TEMPLATE ID's: These numbers and letters are used by programmers and graphic artists when working from the paper version of the script. They identify both a given piece of the script and the graphic template to be used. Template layout specifications contain information regarding the pixel dimensions of the template relative to the main screen.

GRAPHIC: Graphics are described in two ways: first by a boldface identification code, such as **rhintro** (which in this case is shorthand for red hen introduction), then by a verbal description. In some cases, the size of individual graphics is specified in terms of pixels.

TEXT: On-screen text is indicated by the word TEXT followed by the text itself, printed in boldface. For example:

TEXT   **The hen is planting a flower.**

AUDIO: The symbol @ refers to music and sound effects, whereas the letters "im" indicate a separate narrative voice, known as the instructional message. If audio material is not preceded by one of these specific labels, then the dialogue is to be read by the primary narrator.

STUDENT INTERACTION: This section describes the interactive options available on a given screen. It lists the hot spots and tells the programmer what occurs when a given on-screen location is "marked" by the user. (The word mark here is used instead of "click." For example, when the script says "mark the hen," this means "if the user clicks on the hen, the computer will respond as follows.")

In addition to the preceding items, special programming instructions also appear throughout the script, usually in capital letters in response to a user's action. For example, in Vocabulary Screen #2, if the user marks the hen, the script instructs the programmer: DISPLAY LABEL WITH WHITE BACKGROUND FRAME. This means the programmer is to display a graphic label on-screen. (Such labels usually contain text or other information about the object that has just been marked.)

Other programming instructions appear in the form of initials. TO, for example, refers to an action to be generated when the user simply fails to respond. CR refers to an action to be generated after a correct response by the user. The word HELP indicates what happens when the user marks help. Still another form of programmer instructions appears between slash marks, such as /*go to next screen*/.

Following is a list of other codes that are helpful in reading this script:

(!)	a point of emphasis in spoken audio
auwait;	audio pause
ca1	audio when student chooses correct answer on first attempt
ca2	audio when student chooses correct answer on second attempt
em1	audio when student chooses incorrect answer on first attempt
em2	audio when student chooses incorrect answer on second attempt
Correct	correct choice
Distracters	incorrect choice
wa	wrong answer
CR	correct response

To get a feeling for how this production script translates into reality, look again at Part II, the pre-reading activities, at Vocabulary Screen #2, where the educational focus is on action verbs. On this screen, the student will see the Little Red Hen and her animal friends gathered near some flowers in a farmyard. In this picture, the hen is planting a flower, the mouse is watering one, the duck cutting one, and so on. Accompanying this image is the narrator's voice. "Look at the animals," the narrator says. "What are they doing? Mark each animal to find out."

Users then have some options. They can leave the screen and go elsewhere in the program, or they can respond to the narrator and mark one of the animals.

If the student marks the hen, a graphic label appears on the screen, and the narrator describes the action that the hen is taking: "The hen is planting a flower." This audio is followed by the sound of the hen clucking, and then the graphic label disappears. The student then has the opportunity to click on other animals or to go elsewhere in the presentation. Basically, then, reading from top to bottom, the script sequence for each screen indicates the order in which events are to occur.

PART I: From the OPENING SEQUENCE

INTRODUCTORY SCREEN

Script ID: 204a0007h
Template ID: trlrha00
Template Layout: 9.6; 330x390 fullscreen

GRAPHIC: **rhintro**
Size: 330x390
Description: Title screen to story. Picture of the little red hen and the title of the story.

AUDIO @red hen intro music
In this lesson, you will read a story called "The Little Red Hen."

auwait;
AUDIO First, let's find out who is in the story.
DISPLAY FORWARD ARROW

STUDENT INTERACTION
Mark forward arrow:	Student goes on to next scene.
TO:	Use extime for timeout delay; student goes on.
HELP:	Student goes on to next scene.
CR:	Student goes on to next scene.

PART II: From the PRE-READING ACTIVITIES

VOCABULARY ACTIVITIES

OBJECTIVE: To activate prior knowledge of, or develop familiarity with, the vocabulary in the passage.

Vocabulary Screen #2

KEY WORDS & PHRASES: action verbs: plant, water, cut, carry, eat

Script ID: lrhb02
Template Layout: 9.6

GRAPHIC: **rhvoc2**
Graphic Description: The hen, a chick, the duck, the mouse, and the pig are in the farmyard. The hen is planting a flower. The mouse is watering a flower. The duck is cutting a flower. The chick is carrying a flower. The pig is eating a flower. There will be 5 hotspots for sentence labels: one each on the hen, the mouse, the duck, the chick, and the pig.

AUDIO Look at the animals.
What are they doing? Mark each animal to find out.

STUDENT INTERACTION
Hotspots: hen, chick, duck, mouse, pig
It is possible to click on each hotspot many times.

Number of clicks allowed: 30
Forward and backward buttons are functional at the same time as the
hotspots.

Mark forward arrow:	Student goes on to passage presentation.
Mark backward arrow:	Student goes back to previous screen.
TO/HELP/CR:	Display label/text/audio for each animal.
	Then go on to passage presentation.

Mark the hen:
DISPLAY LABEL WITH WHITE BACKGROUND FRAME
TEXT   **The hen is planting a flower.**
AUDIO   The hen is planting a flower.
           @hen clucking

auwait;
REMOVE LABEL

Mark the mouse:
DISPLAY LABEL WITH WHITE BACKGROUND FRAME
TEXT   **The mouse is watering a flower.**
AUDIO   The mouse is watering a flower.
           @mouse squeaking

auwait;
REMOVE LABEL

Mark the duck:
DISPLAY LABEL WITH WHITE BACKGROUND FRAME
TEXT   **The duck is cutting a flower.**
AUDIO   The duck is cutting a flower.
@duck quacking

auwait;
REMOVE LABEL

Mark the pig:
DISPLAY LABEL WITH WHITE BACKGROUND FRAME
TEXT   **The pig is eating an apple.**
AUDIO   The pig is eating an apple.
           @pig oinking

auwait;
REMOVE LABEL

Mark the chick:
DISPLAY LABEL WITH WHITE BACKGROUND FRAME
TEXT   **The chick is carrying a flower.**
AUDIO   The chick is carrying a flower.
           @chick peeping

auwait;
REMOVE LABEL

PART III: From the INTERACTIVE READ

Scene 1

GRAPHIC: **rhlived**
Graphic Description: The little red hen and her chicks are on the farm where they live.

DRAW BACKWARD AND FORWARD ARROWS: **frwrd1, bckwrd1**

/*note to programmer: put small headphones at the beginning of the sentence*/

TEXT	**Once upon a time, a little red hen lived with her chicks on a farm.**
AUDIO	Once upon a time, a little red hen lived with her chicks on a farm.

STUDENT INTERACTION

Mark forward arrow:	Student goes on to next scene.
Mark backward arrow:	Student goes back to previous scene.
Mark Audio Repeat:	Student hears last audio instructions.
Mark a word:	Student hears a word.
Mark small headphones:	Student hears sentence.
TO:	extime variable tied to passage presentation time; student goes on.
HELP:	Student goes on to next scene.
CR:	Student goes on to next scene.

Scene 2

GRAPHIC: **rhfound**
Graphic Description: The little red hen is in the farmyard. She has just found some grains of wheat. She is bending over and picking up some of the grains. The expression on her face shows that she is happy.

DRAW BACKWARD AND FORWARD ARROWS

/*note to programmer: put small headphones at the beginning of the sentence*/

TEXT	**One spring day, the little red hen found some grains of wheat in the farmyard.**
AUDIO	One spring day, the little red hen found some grains of wheat in the farmyard.

STUDENT INTERACTION

Mark forward arrow:	Student goes on to next scene.
Mark backward arrow:	Student goes back to previous scene.
Mark Audio Repeat:	Student hears last audio instructions.
Mark a word:	Student hears a word.
Mark small headphones:	Student hears sentence.

TO:	extime variable tied to passage presentation time; student goes on.
HELP:	Student goes on to next scene.
CR:	Student goes on to next scene.

PART IV: From the COMPREHENSION EXERCISES

Primary Exercise 1.1

SCRIPT ID: Irhd0108

GRAPHIC: **rhhenvoc**

TEXT ANSWER CHOICES
Correct Answer: **hen**
Distracters: **chick, pig**

AUDIO
im: Oh, look! A hen.
Move the word hen to the empty box.

FEEDBACK MESSAGES
ca1:   You moved the word hen.
ca2:   Now you moved the word hen.
em1:   Try again. That is the word _ [wa]. Move the word hen.
em2:   That is the word _ [wa]. Here (!) is the word hen.
TO1:   Oh, look! Do you see the word hen? Move the word hen.
TO2:   Look! Here (!) is the word hen.
HELP: Look! Here (!) is the word hen.

Primary Exercise 2.1

SCRIPT ID: Irhe0101

GRAPHIC: **rhfndwht**

AUDIO   What did the little red hen find in the farmyard?

TEXT ANSWER CHOICES
Correct: **some grains of wheat**
Distracters: **some bags of flour**
some loaves of bread

AUDIO ANSWER CHOICES
Correct: some grains of wheat
Distracters: some bags of flour
some loaves of bread

im: Mark your answer.

FEEDBACK MESSAGES
ca1:   She found some grains of wheat.

ca2: She found some grains of wheat.
em1: She didn't find [wa] . What did (!) she find?
wa2: She didn't find [wa] . She found some grains of wheat.
      /*grains of wheat is highlighted*/
TO1: What did the little red hen find? Mark your answer.
TO2: The little red hen found some grains of wheat.
      /*grains of wheat is highlighted*/
HELP: Same as for TO2.

Primary Exercise Set 3

Objective: Recall sequence of events in a reading passage.

Template Layout: 7.1

GRAPHICS LIST FOR THIS SET:
rhfndwht
rhpinwht
rhwtrwht
rhcutwht
rhcarwht
rhmadbrd

Primary Exercise 3.1

SCRIPT ID: Irhf0102

GRAPHIC: **rhfndwht**

AUDIO   The little red hen found some grains of wheat in the
                  farmyard. What did she do next?

TEXT ANSWER CHOICES
Correct: **She planted the wheat next.**
Distracters: **She watered the wheat next.**
                  She cut the wheat next.

AUDIO ANSWER CHOICES
Correct: She planted the wheat next.
Distracters: She watered the wheat next.
                  She cut the wheat next.

im: Mark your answer.

FEEDBACK MESSAGES
ca1: She planted the wheat next.
ca2: She planted the wheat next.
em1: Think again. The little red hen found some grains of wheat. Then
      what did she do?
em2: She didn't (water/cut) (!) the wheat next. She planted (!) it.

TO1:   Look! The little red hen found some grains of wheat. Then what
        did she do?
TO2:   She planted the wheat next.
HELP: She planted the wheat next.

Primary Exercise 4.3

SCRIPT ID: lrhg0305

GRAPHIC: **rheatbrd**

AUDIO
im: How did the little red hen feel when she ate the bread?

TEXT ANSWER CHOICES
Correct: **She felt happy.**
Distracters: **She felt sad.**
                        She felt tired.

AUDIO ANSWER CHOICES
Correct: She felt happy.
Distracters: She felt sad.
                        She felt tired.

AUDIO
im: Mark your answer.

FEEDBACK MESSAGES
ca1:   She felt happy when she ate the bread.
ca2:   The little red hen felt happy.
wa1:   The little red hen felt happy. Mark the happy (!) face.
wa2:   Look at the little red hen. She felt happy (!) when she ate the bread.
TO1:   The little red hen felt happy. Mark the happy (!) face.
TO2:   Look at the little red hen. She felt happy (!) when she ate the bread.
HELP: Look at the little red hen. She felt happy (!) when she ate the bread.

PART V: From the CLOSING ACTIVITIES

CLOSURE/WRAP-UP

Activity 1

SCRIPT ID: lrhw01

GRAPHICS: **rhendwrp, rhyes, rhno**

AUDIO   @le red hen transition music
                        Did you like the story of the little red hen?

BUTTONS
YES   NO
Audio feedback for Yes: Great! This is (!) a wonderful story.

Audio feedback for No: Oh, well. Maybe you'll like the next (!) story.
HELP/TO: Hope you liked the story.

Activity 2

SCRIPT ID: Irhw02

GRAPHICS: **rhendwrp, rhyes, rhno**

AUDIO   Would you like to read the story of the little red hen again?

BUTTONS
YES   NO
Audio feedback for Yes: Okay! Let's read the story again.
Audio feedback for No: Okay! Let's go to the notebook.
HELP/TO: Let's go to the notebook.

STUDENT INTERACTION
Mark Yes: Speak audio feedback, above. Go to Opening.
Mark Forward Arrow: Go to Notebook activity.
Mark No: Speak audio feedback, above. Go to Notebook activity.

C

Web Site Specifications for the Apple Personalized Internet Launcher

The following is an excerpt from a concept document describing a Web site produced by CKS Interactive for Apple Computer Company. The Web site is intended for users of Apple's Performa Computers as a way of familiarizing them with the Internet, and to build product loyalty as well.

The document that follows can be divided into three parts: a project description, a content overview, and a delineation of the user's experience. Also included is a modified version of the original flowchart.

Documents such as this are conceptual in nature and offer a starting point for both the design team and the client who commissioned the project. As the project evolves and the design process continues, specifics of content and layout may change, usually as the result of a continuing dialogue between the client and the design team.

I: PROJECT DESCRIPTION

Objectives

There are five objectives of the Apple Personalized Internet Launcher (PIL) Web site:

- Provide a vehicle for the Apple Americas Consumer Division to get closer to their customers and receive feedback from customers in order to create relevant marketing and product solutions.

- Encourage add-on purchases of Apple and ISV (Independent Software Vendor) products.

- Expand Apple's enthusiastic customer base and convert them into proactive evangelists.

- Provide Apple customers with a customized tool that provides easy access to relevant information on the Internet and Apple product information.

- To improve and maintain customers' positive attitudes and loyalty toward Apple.

Target User

The primary targeted users for the PIL Web site consist of recent purchasers of Performa computers. The secondary targeted users for the PIL Web site consist of existing owners of Macintosh computers as well as new purchasers who do not purchase Performas.

Recent purchasers of Performa computers range in abilities. Many are novice users who are buying a computer for the very first time. They may be intimidated by computer-speak, fearful of making mistakes, and had to overcome major barriers in order to make the choice of buying a Macintosh. Some still need to learn to use a mouse. The other half have purchased before, most likely a Mac, and are upgrading. The PIL Web site will be designed to make things easy enough for the first-time user, but not limit the more advanced user.

Description of the PIL Site

The Apple PIL will appear as two different sites; we refer to them in this document as the "Premium" site and the "Standard" site. They will have the same URL.

- The Premium site will be optimized for users with modern browsers— those capable of supporting the Persistent Client State HTTP Cookie mechanism. Browsers supporting this standard today include Netscape Navigator 1.1 and later, as well as Microsoft Internet Explorer 2.0 and later.

- The Standard site will be optimized for users with older browsers which are not cookie capable, and for those users who do not care to register.

II: CONTENT AREAS

The Premium Site

The intent of the Premium site is to meet all of the objectives outlined earlier in this document, offering the novice and more advanced user their own personalized Web site with an easy way to understand and access relevant information on the Internet and Apple product information.

Content Area 1. The Home Page Area

This area [see Figure C.1] will include:

- a personalized welcome message
- support for multiple users
- an area for the user's top twelve personally selected Web sites
- a search engine
- access to personalized suggested Web site categories
- access to a sites directory
- access to Apple and ISV product information
- access to Apple technical support
- access to Apple questionnaire (Tell Apple . . .)
- access to a basic tour
- access to on-line registration
- access to the ability to further customize their PIL

Note: Under the User Experience portion of this document in the Design Guidelines and Tone of Copy section, it is recommended that the home page not be cluttered but should be simple and straightforward in keeping with Appleesque design. Given the current number of items specified to be on the home page, this is a daunting task for the CKS design team. To that end, the team is looking at design options which will meet Apple marketing goals but simplify the home page to meet Apple design guidelines.

Content Area 2: Registration, Basic Tour, Customization of Page

This area [see Figure C.1] will register, educate, and offer the users further customization of their PIL site. The registration area will include questions to gather the information needed for CKS to serve up a PIL to the user, as well as demographic and gender information potentially needed to further customize the page in the future.

The tour will include basic information on the user's PIL, key features in a cookie capable browser, and the Internet. It will focus on features which might pose some difficulty for a novice user and aim to educate the user.

REGISTERED USER'S TOUR: MAIN TOUR PAGE

What's available on the Main Tour page of the Registered User's Tour:

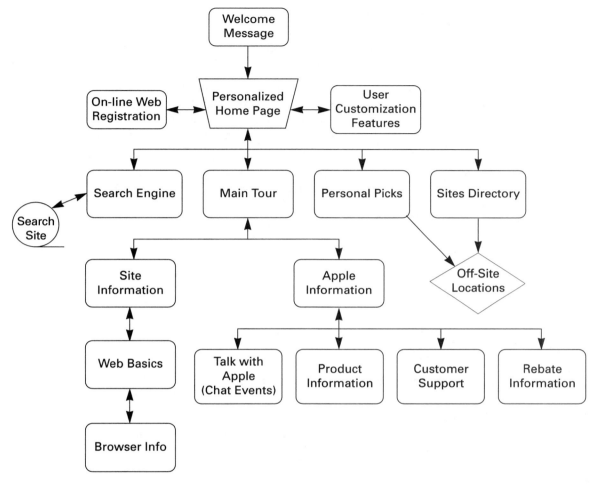

Figure C.1
Web Site Architecture Flowchart. This chart maps out the underlying interactive structure for the Apple Web site described in this appendix.

- basics on uniqueness of the personalized home, motivation for registering

- a visual layout of the home page in full view (no scroll bars), with labeled components (My Home, Suggested Categories, Apple Sites, My Favorite Site)

- link to information about the PIL Site and description

- link to Web Basics information and description

- link to Browser Information and description

Content Area 3: The Suggested Sites

This area will include a subset of the sites directory that offers users categories based on their registration input either during online registration on their Performa or when they first arrive at the PIL site. Each category will generate a list of six preselected Web sites from which a user can select, review, and then go to the site. These six sites will change on a daily basis. All of the personalized categories and the site directory should be accessible from each of the individual category pages. The sites directory should take the user to the area in the directory related to the personalized category; i.e., if the user is in a personalized music category page looking at their six sites of the day, then they can select the site directory which will take them to the expanded list of music sites in the site directory.

Content Area 4: Apple Info

Content Area 4 [see Figure C.1] will include a 2-way dialog with Apple and Apple information:

- a 2-way dialog with Apple (Talk with Apple . . .) close-ended questionnaire—user forum monthly chat events

- Apple and ISV product information

- Support—support bulletin board

- rebates & rewards

Content Area 5: The Sites Directory

The generic sites directory [see Figure C.1] will be accessible from the home page as well as each of the four individual personalized categories generated in suggested sites. The home page will provide access to a specific subset of the categories offered in the sites directory database, based on personalization data.

The home page will also provide access to the sites directory for searches beyond the personal categories. Users will be able to customize their sites to conduct searches in 20 different lifestyle categories:

Arts & Humanities
Computing
Food & Drink
Games
Government & Politics
Health & Fitness
Home Hobbies
Home Learning
Internet References
Movies
Music
News
Parenting

Personal Finance
Religion & Philosophy
Science
Shopping
Sports & Recreation
Travel
TV & Radio

III: USER EXPERIENCE ELEMENTS

1. <u>Content</u>: An overview of the content appears on the preceding pages. This content will be further developed, made specific, and integrated into the user experience as described below.

2. <u>Site Organization</u>: The Apple Personalized Internet Launcher (PIL) is actually a series of hierarchically arranged pages. The first page the user sees, myhome.apple.com, acts as the main page.

This main page offers the user:

- access to three different kinds of links to other places on the Web: those suggested by the PIL, those they personally select, and some offered by Apple

- buttons offering direct links to suggested categories thought to be of highest interest, based on the personal profile gathered from on-line registration

- the ability to create additional personalized home pages for multiple users

- basic WWW search functionality, linked to a public search engine

- functionality for further customizing the home page, including a way to access on-line registration or a personalized profile to generate the data for personalization (and recreate the home page)

- the presentation of some dynamic and provocative information that is new and different every time the user comes to the page, such as a news or business update or information about other cool Web sites

- access to some introductory/overview information

- advertising space to direct users to functionality on the site or elsewhere on the Web

Other pages include:

- Suggested Sites Page
  Links to a manageable number (20–30) of sites automatically generated by a database, based on information in the user's registration profile, and updated regularly to include current information on new, cool or interesting Web sites. These site links are broken into 4 to 5 categories of sites, obtained directly from the profile, with 4–6 links in each category.

- My Top Twelve Sites Page
  Links personally selected by the user for direct access to other Web sites, updateable on demand.

- Apple Sites Page
  Links to standard Apple-provided features, including:

  Tell Apple: A close-ended questionnaire with the ability to submit input directly to Apple. Submitting a questionnaire will generate and send a thank-you letter to the user.

  Product Information: 10% listed and 90% linked information about Apple and third-party hardware and software products, as well as advertising space for 2 ads, highlights of developers who have rebates that month, and a link to a software rebate.

3. <u>Site Navigational Scheme</u>: Users will move from page to page in the site by selecting from the local links available on the home page. Thus, moving to another page within the PIL will appear quite similar to moving to another page at another site on the Web. This will not only help to unify the interface, but teach users about the general experience of using the Web, since this is the very same behavior exhibited at any other Web site. A universal tool bar should be included on all PIL pages, enabling users to easily move within the PIL site.

Links will be listed with visuals (icons or graphics) and/or the conventional hypertext. Typically, the first few links on each of the Suggested Sites pages and My Top Twelve Sites area will have both visualized and underlined text. This way, novice users will be able to click on understandable graphics, and more savvy Web users can use the underlined links to go directly to a desired location. This method will also educate novice users about how links are listed at other sites on the Web, since though they may start using only the visual links, they should quickly learn the semantics of the underlined text links.

A critical navigation method for the Netscape browser is the Bookmarks menu, and similar functionality will be incorporated in the Apple PIL. Novice users will need to learn about the Bookmarking features and how to use them. The items in the My Favorite Sites menu for the Personalized Internet Launcher will complement the links on the page, and include a small set of links useful to the customer, including "My Home" and the Apple corporate Web site. The "My Home" link will be the first menu item in the Bookmarks file. Information about Bookmarks will also be available from the information link on the home page.

4. <u>Design Guidelines and Tone of Copy</u>: The Apple Personalized Internet Launcher should be inviting to novice users, simple and straightforward, and yet encourage customers to return to this site and to use it as a gateway to the World Wide Web and Apple. It should also be Appleesque, include Apple brand mark and logo type on all PIL pages, clearly reflect the Apple culture and emphasize that Apple is making the access of useful sites on the Internet possible. The home page should thus have a

clean, somewhat futuristic visual treatment, most probably with a white background, little clutter and with Apple imagery subtly incorporated. The visuals of the links themselves should demand primary attention, to draw users directly to the Internet functionality.

Still, the aesthetic must be appealing enough so as not to bore or offend more advanced users. Users might be able to customize the visual design to further promote the idea that this is their very own home page. Customizations could include choosing from a number of preconceived designs, adjusting colors or textures, or including certain visual elements at the user's request.

The terminology used throughout the site will be friendly and inviting. The division of the links into Suggested Sites and Sites Directory should help users to understand the difference and why the two types would be valuable to them.

At the same time, the verbiage should take advantage of some actual Web terms to help educate users, but not those terms that are so technical that they might confuse users. Employing terms that users will need to know, such as link, site, and page, will help users to grow more comfortable with the Web. Verbiage such as URL, HTML, and other more technical terms should be used with caution—we don't want to shield users from them altogether, but should be careful to introduce them through well-defined usage.

5. <u>Consistency of Look and Feel</u>: The Apple Personalized Internet Launcher site will have a unified visual and navigation scheme to ensure a consistent look and feel. All pages in the site will have a similar general treatment, so that once users have seen the first personalized page, they will know when moving among all pages in the site that they are in the personalized home page site.

As with many Web sites, the PIL will have a standard array of buttons for navigation to local pages at the bottom of each of the screens. These will include My Home, Suggested Sites, My Sites, and Apple Sites.

If users are able to customize the visual design in some way (as noted above), the customized design must be the same across all pages at the site to ensure consistency.

6. <u>Assumed User Platform and Incoming Bps Rate</u>: The Premium and Standard sites will be optimized for users with Netscape Navigator 2.0 and designed for the minimum Macintosh system requirements recommended to support Netscape Navigator 2.0. The page size for the PIL site will be designed to the default size that Netscape Navigator opens up at on the Mac platform.

- 68020 processor

- 8 MB RAM

- 2 MB of disk space available

- Macintosh system 7 or later and Power PC platforms are supported

- monitor NOT specified

- modem NOT specified

7. <u>Use of New Technologies</u>: The Premium site may make use of frames in Netscape 2.0 and other new Web technology as it is available, depending on final design directions.

8. <u>Level of Personalization</u>: Another important aspect of the Apple PIL is that it personalizes the Web to individual needs, making the Web less overwhelming to novices and experts alike. Personalization is achieved in a number of ways.

The user's name, gathered from the registration process, will be used in dialog throughout the home page site to help the site seem more personal. The user will be greeted on the home page by name, and both the Suggested Sites and My Sites pages will incorporate the user's name (though the links to get to these pages from the home page will not). The use of names will help a user on a multiple-user machine verify that he is on his own home page and not another user's. Names will not be overused, however, and may not be appropriate, for instance, on every page.

D

Excerpts from the Treatment of the Adventure Game *Dreggs*

Dreggs is an interactive adventure game in which the player takes on the role of prophesied messiah. The world of *Dreggs* is a world out of balance, a world inhabited by dragons, poachers, demons, and fallen tribespeople. It is a once-advanced civilization that was destroyed by the greed of its inhabitants. The player's goal is to set the balance right again. In order to do this, the player must navigate throughout various aspects of this world, including time.

Dreggs takes advantage of live-action photography and computer animation to infuse the traditional game environment with narrative elements and scenic displays more often associated with film and television. It was written, directed, and produced by John Kuri of Fathom Studios in Sausalito, California.

Included here are excerpts from the treatment, a document used in this instance to establish the fundamental elements of both the gaming and narrative experiences.

THE ENVIRONMENT

Dreggs is set within the milieu of a lost civilization that has contained itself inside the labyrinth of a mountain and the body of water harnessed by it. The key settings are:

<u>Chronus Realm Chamber</u>: Massive in scale, this cavernous chamber actually is the nucleus of the labyrinth which makes up Dreggs. Seven passageways radiate from the chamber's perimeter. Huge hieroglyphics are faintly in view, parts of them missing from years of erosion. In the center a shaft of hazy light finds its way through an opening from

above, falling upon an area of the rocky floor revealing a metallic object that penetrates from the rocks. It is the Chronus Realm, time key to all of Dreggs.

The Chronus Realm: A semi-spherical dome in an elliptical shape resembling an egg. It is bronze in color, with a gnomon, its relief casting a shadow across the outer perimeter of the Realm. Within the center is the face of a female, the gnomon protruding from the bridge of her nose. Around the perimeter are hieroglyphics contained within specific degrees of the sphere. As the Realm is viewed the shadow cast by the gnomon will move, like a sundial, casting a dark line across a symbol.

A narrative in a woman's voice will become very audible, as long as the Player remains close to the Realm. The narration will relate to some aspect of the hieroglyphic being highlighted by the shadow line and will be heard as long as the Player remains in that spot. During game play, as the Player returns to the Realm the time will be different, thus the shadow is cast upon a different hieroglyphic, and thus the narration will be different. The Player will realize that a living history of Dreggs is being spoken but also that the Chronus Realm and the light moving across its face represent a key to accessing the network of corridors that connect Dreggs.

The Septentrion: An outer circular corridor which surrounds the Chronus Realm Chamber and interconnects the seven passageways into the world of Dreggs. The Septentrion revolves around the chamber, and in doing so only allows complete inner connection at specific moments in time. Its rotation is controlled by the light passing across the Chronus Realm.

The Abyss: The dark inner world of Dreggs. Its inhabitants, Poachers, live within its damp, sooty environment, in hovels made from eggshell halves connected to each other in a honeycomb-like arrangement. In the center of this chamber is an island surrounded by dark, contaminated water. On the island is a throne made from the bones, talons, and skins of dragons. From here a god-like voice of doom issues a dark manifesto which seems to have control over the Poachers. The throne is protected by Cerberus, a three-headed monster.

The ambient sound of the Abyss will emphasize the mechanical/slave-like condition which will be evident in the attitude of its inhabitants. Music, a la Jimi Hendrix's "Purple Haze," will underscore the feeling evoked from the Abyss.

The Player will discover young water dragons corralled in slave labor along the perimeter of the island, drawing dark water through their mouths, filtering it and then purging it into containers made from eggshells. Poachers carry away the water containers. Young earth dragons use their tails like augers, drilling through the walls of dark soil, extracting a coal-like material, which is loaded into awaiting eggshells and carried away by Poachers. The eggshells filled with coal arrive at a furnace area where young fire dragons breathe flames on a huge pit of

Figure D.1
Dreggs: The Binary
Pulsar. The story
experienced in
Dreggs takes place
in the present day.
The Player will
discover life on the
binary twin of Earth.

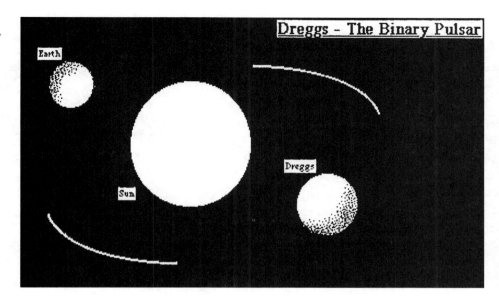

Figure D.1
Dreggs: The Binary Pulsar. The story experienced in *Dreggs* takes place in the present day. The Player will discover life on the binary twin of Earth.

coal, igniting the coal, which creates a dim heat along with tons of soot which hangs in the air.

<u>The Nest</u>: An ideal environment for the incubation and hatching of the dragon eggs. It is located near the uppermost area, along the edge of Lethe, the great body of water [see Figure D.2]. Being a shallow cave, it is open in the direction of the lake but has a passageway that leads to the interior of the mountain.

<u>Lethe</u>: A massive lake, held within its confines by the mountain. Hydraulic pressure forces water into the different layers of Dreggs creating rivers cut from time that allow another means of navigating this world.

<u>The Passageways</u>: The myriad of rocky corridors which interlink the Chronus Realm Chamber, the Abyss [see Figure D.3], the Nest, and Lethe.

<u>A general note as to the visual aesthetics</u>: In the creation of the backgrounds used in Dreggs, we will use photographic elements taken from natural environments in conjunction with artistic illustrations. The visual effect will be an essential and intriguing aspect of the experience.

THE MAIN CHARACTERS

<u>The Player</u>: The experience of this game will be in the first person. Therefore, the characters discovered along the journey will interact with the Player in dialogue and physical attitude, making the Player active in the outcome of the characters' movements.

Figure D.2
Dreggs Section
View: Overview
and Level 1.

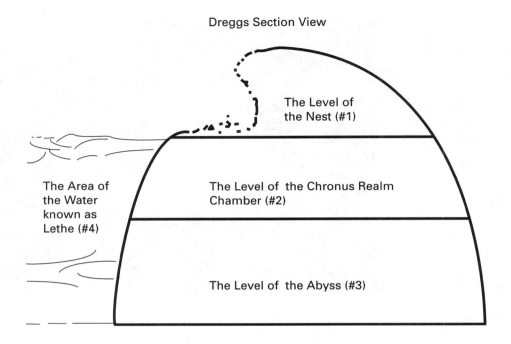

Dreggs Section View

The Level of
the Nest (#1)

The Area of
the Water
known as
Lethe (#4)

The Level of the Chronus Realm
Chamber (#2)

The Level of the Abyss (#3)

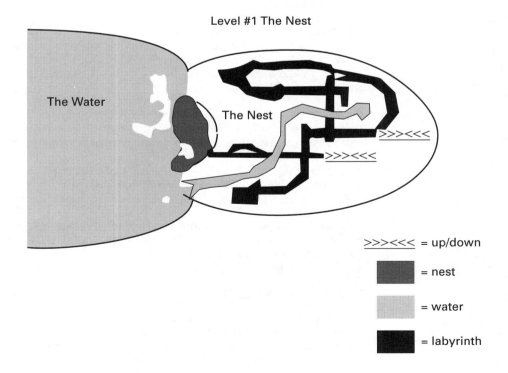

Level #1 The Nest

The Water

The Nest

>>><<< = up/down

= nest

= water

= labyrinth

Figure D.3
Dreggs Section
View: Levels 2 and 3.

Level #2 The Chamber of the Chronus Realm

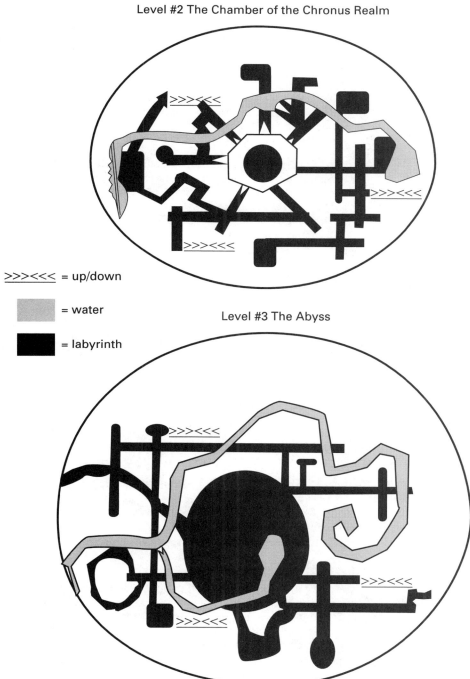

>>><<< = up/down

▨ = water

■ = labyrinth

Level #3 The Abyss

The High Priestess: A live-action actor, she will be both a help and hindrance to the Player. She is a dual-personality, schizophrenic woman of extraordinary beauty who is in an eternal struggle between the forces of light and dark. She has enormous powers of illusion and uses them freely. The Player's actions will have a direct effect on the appearance and condition of the High Priestess. Positive actions will cause the High Priestess's beauty to be unmasked while negative actions will cause her mask to become more complex and intimidating. Her mask will be applied in multiple layers, any or all of them being used to illustrate her various states of health.

The Poachers: The creatures of the Abyss; their true faces are covered by masks which project peculiar characteristics. They do negative work for the High Priestess, stealing dragon eggs and transporting them to the Abyss. They worship the dragon eggs because they represent all that life is. Evidence of this is seen in the use of the eggshells as building material within the Abyss. The Poachers recognize the Player as a threat.

The Creatures: Human-like tribal characters who inhabit some of the upper regions. They recognize the Player as the means to the fulfillment of the promise for new life. Their numbers have been greatly reduced since many have gone over to join the Poachers and serve the High Priestess.

The Mother Dragon: The embodiment of all the powers of Mother Nature, she will deposit her eggs in the Nest. This new life will bring a regeneration of the four primary elements, Earth, Air, Water, and Fire to the planet. She is immense in size and power and is capable of unleashing her fury on this world or rewarding it with her gentleness.

The Eggs: Amidst the numerous eggs laid by the Mother Dragon are four very precious ones which contain the power of each of the elements. When they hatch they will produce four distinct dragons, representing Air, Water, Earth, and Fire. The features of these dragons [illustrated in original treatment] are drawn from the elements they represent.

The Chronus Realm: As discussed above, it is the teller of the living history. It will provide the Player with information that is key to unlocking the secrets of Dreggs. The Chronus Realm also represents time, an essential character within Dreggs, and thus the Chronus Realm becomes the axis around which all things happen.

INTERACTIVITY

Time as an Interactive Element: The Player will determine that there exists a connection between the Chronus Realm and the rotation of the Septentrion and thus the accessibility of different passageways within Dreggs. However, players will also discover that the location of their entry point in the game will vary. The game will be a multithread story programmed so that action unfolds without regard to the Player's location. The Player will be able to experience the ongoing story from different locations or within the perspective of different characters. Each thread of the story is complete within itself, but the other threads will remain for the Player to experience.

The Labyrinth as an Interactive Element: The rotational aspect of the Septentrion will create an ever-changing environment for the Player. At one moment a passageway may lead to or from the Chronus Realm Chamber. In another moment the Player may find that passageway blocked. The interconnectivity of the levels will be effected in a similar manner, all of which will take away the Player's ability to memorize the roadmap.

Sound as an Interactive Element: Directional sound will be used to further enhance game play. The Player will be given sound effects which will cause a reaction, be it curiosity, jeopardy, or bewilderment. The Player may try to avoid the source of the sound or in some cases be drawn to it in order to inspect the source. The characters with whom the Player interacts will be reactive to the sound effects, further enhancing their use.

The Player Interface: The Player experiences this story in the first person. The interface will be through the use of mouse, joystick, or keyboard directional arrows and space bar. It is our intent to truly make this a first-person perspective in the same manner that a professional flight simulator presents only the view from the cockpit. The entire experience will be through the Player's point of view.

The Player Threats: Traveling through the labyrinth the Player will encounter multiple threats. These will be a combination of environmental elements (quicksand, earthquakes, floods, extreme heat, extreme cold, polluted air, heavy winds, lava, swamps, etc.). As well, the Player will encounter the threat of the Poachers, who view the Player as the one who will steal their power source. The Player will find illusions presented at random by the High Priestess who will present tests of various forms and magnitudes. Failure could bring an end to the Player.

THE INTRODUCTION OF PLAY

The screen is black. The sound of rock being chipped away. The screen pulses as if coming to life.

A jagged band of light appears across the screen. The rough edges of light seem to animate as we hear more hammering, now accompanied by labored breathing.

Suddenly, the crack of light gives way to a full-screen image as the shell falls away, revealing full screen the very strange face of a Poacher, a bony long-limbed creature. It is leaning over, eyes piercing through the holes in a bizarre mask which imitates the features of a human. Its bony finger tentatively probes toward the screen, almost seeming to penetrate to the Player. But it suddenly withdraws and turns as the Player's POV (point of view) begins to rise out of the shell. The Poacher runs out of the rocky chamber.

By moving the mouse, joystick, or directional arrows, the Player will explore the Chronus Realm Chamber. Within the walls of this rocky chamber will be the skeleton of a large dragon. The scale of this chamber will be overwhelming, the dragon's skeleton forty feet long and only partially revealed, having been covered for years by rock and rubble that have fallen over its lifeless remains.

In the Chronus Realm the Player will learn that the Mother Dragon has deposited eggs in a nest high up in the mountain. In a hologram-like image, the High Priestess will reveal herself, urging the Player to find and guard the eggs.

The Player will find comradeship with the few Creatures who remain in the upper regions. They will meet and join forces with the Player as the passageways are navigated. The Creatures worship the Dragon Eggs for the right reasons and see the Player as the one who will bring forth new life. The Player, accompanied by a Creature, finds the dragon's nest, but to the Creature's horror, Poachers have stolen most of the eggs.

Mother Dragon discovers what's happened and unleashes her fury on Dreggs. In relation to the environment she's the size of the mother ship in "Close Encounters," and the visualization of her fury will be a spectacle in itself. She turns into the equivalent of the Four Horsemen of the Apocalypse as she lets loose a cry of painful devastation followed by a firestorm from her mouth and a devastating wind from her wings.

And thus the adventure begins. The Player must stop the Poachers and bring the four primary eggs back to their nest. If the Player succeeds, Mother Dragon will stop her fury and return her gentleness to the planet. But to accomplish the task the Player will encounter all that is contained within this world.

GLOSSARY

Adventure game   Computer role-playing game in which the player takes on a persona engaged in a quest. Skills involved may include those of the traditional arcade, as well as solving puzzles, riddles, and other problems.

Alpha test   The first test by an audience of a given piece of computer software. Alpha tests are used at an early stage to help catch design and content flaws.

Arcade game   Computer game in which game play revolves around the user's dexterity and ability to respond rapidly to the movement of objects on screen. These games have their roots in the old arcade games found in carnivals.

Assessment   Term used by instructional designers to indicate tests and testing procedures employed to evaluate student progress in an educational context.

Associative logic   Nonlinear or nonrational thought patterns. One example is the human mind as it wanders from memory to memory after being triggered into reverie by apparently random associations between present surroundings and past events. In associative logic, the connections between ideas are based on sensual similarities, such as color or shape or smell, rather than on traditional syllogistic logic.

Avatar   The visual representation of a fictional persona used by chat room participants and on-line game players.

Bandwidth   Term used to describe the capability of wiring systems to deliver media elements from the source to the computer screen. Generally, the greater the bandwidth, the more information that can be delivered at any given instant.

Beta test   Second stage testing of computer software, done after any design changes of the alpha prototype have been implemented.

Blue screen   Empty backdrop that is placed behind studio actors to serve as a composite screen for the electronic construction of digital backgrounds.

CD-ROM   Compact disk with read only memory. A medium capable of holding video, audio, and text

elements that can be played back and viewed on a computer-controlled monitor.

Chat rooms   On-line sites where participants can partake in simultaneous, two-way dialogue with others using the computer keyboard to type messages that are transmitted instantly. Recent chat rooms include avatars in a graphic environment, thereby adding a visual dimension to the conversation.

Classification scheme   Method of design in which material is broken into categories that are presented on screen.

Collage   Method of creation using juxtaposition of elements.

Comparison/contrast   Form of classification in which two or more subjects are broken down into categories and analyzed according to similarities and differences.

Concept development   The first stage of the interactive design process; in it, the design team works to define and develop the underlying concept of an interactive program.

Conditional linking   Programmed setup of hyperlinks that are variable-dependent and therefore capable of routing users in different ways, depending on remote as well as immediate circumstances.

Content expert   Consultant or staff person who serves as an expert concerning a particular body of content.

Corporate training   Educational programs instituted by businesses and other institutions to increase sales, productivity, or morale. These programs often follow an instructional design model and involve the use of traditional as well as interactive media.

Decision point   Place in an interactive pathway (or on a given screen) in which users must make a decision that will move them in one direction while eliminating other directions as options. Also referred to as nodes.

Design document   Broad term most often used to describe a document that contains visual and programming specifications for interactive media. Sometimes used more generally to describe the production script in particular, or even an earlier-stage conceptual design.

Design metaphor   Method of design in which material is conceptualized on screen by means of visual metaphors that make use of inherent underlying similarities between the material and some object to which it is being compared.

Design shell   Conceptual notion useful in concept development for interactive media. In a well-constructed design shell, three elements must work together in concert: content, interface, and interactive structure.

DVD   Digital video disk. Format capable of delivering more video of higher quality than CD-ROM.

E-mail   Electronic mail written on a computer terminal and delivered via modem. It is generally stored on a server until accessed by the intended recipient.

Energeia   Aristotelian term used to describe the inner energy or stage of being that compels a fictional character forward through a sequence of events. Described by novelist John Gardner as "the actualization of the potential which exists in character and situation."

Event metaphor   Design metaphor in which the structure of an interactive program is organized around the various stages of some specified event.

Exploratory programs   Type of educational program in which the emphasis is on allowing the students to explore a subject matter, create artifacts, and run experiments.

Expository structure   Term used to describe the shape of a nonfiction presentation intended to explain or persuade. Constituent elements include introduction, division, background, main body, and conclusion.

Flowcharts   Diagrams used by writers and interactive designers to help visually conceptualize a program's flow and structure.

Format   Software or delivery media used to play and carry content on a given computer platform.

Foundational programs   Educational programs that teach basic skills in a particular subject matter or discipline. The focus of foundational software is on quantifiable skills.

Functionality   Term used to indicate practical matters regarding the function of the program. In a written script, the portions dealing with functionality specify the possible movement between screens, the linkages, and other navigational and situational elements.

The Futurists   Group of writers and artists who at the turn of the twentieth century sought to revolutionize society through a fusion of the arts and the sciences.

Guide figure   Character whose reoccurring presence serves the function of guiding users' through the program.

Hierarchical pattern   Interactive pattern in which the screen-to-screen movement proceeds from category to subcategory on the basis of users' choices.

Hot spot   A place on the screen that triggers a media event when activated by a user. The event might be a display of text, sound, or image.

HTML   Hypertext Mark-up Language. A programming language used by writers working on the Web to indicate page breaks, build links, and insert graphics.

Hyperlinking   A method for moving a user between related ideas. Hyperlinks can trigger animations or other media events without transporting the user to a new screen, or they may move the user to a new screen location altogether.

Immersive environment   A life-like environment that simulates a real-life experience or place and requires users to play an extended role within the simulation.

Instructional design   The design of training or educational material according to a systematic approach in which the presentation of concepts and material is followed by student activities, review, and assessment. (See also *Instructional goals* and *Needs analysis.*)

Instructional goals   The statement of instructional goals comes at the beginning of the instructional design process, after the needs of the audience have been defined. It involves the statement of specific educational objectives, including skills and concepts to be learned by students in a particular course. (See also *Instructional design.*)

Interactive brochure   Advertising that is delivered via disk or CD-ROM and features interactivity and the possible use of graphics, sound, and/or video.

Interactive cinema   The merging of traditional cinematic concerns with computer gaming in an interactive environment. Interactive cinema has much in common with adventure games, but story and character concerns tend to be more dominant.

Interactive structure   The overall shape and design of the information pathways that underlie a program.

Interactivity   Conceptual notion that gives users selective access and control over material. In interactive media, users have an increased role in structuring their experience.

Interface   The place where the computer and the user meet. Term used to describe the graphic display of interactive pathways on screen, including placement of navigational tools, icons, and information categories.

Internet   Interconnected public network of institutional servers, wiring systems, and personal computers that allows electronic communication of text, audio, and video and the transfer and storage of digital information.

Intranet   Secured, private network that provides direct access from selected institutional servers to personal computers via modem. Used to provide enhanced media capabilities and transmission speeds that are often unavailable over the Internet.

ISDN line   High-bandwidth wiring system capable of delivering enhanced multimedia to businesses and homes. It is superior to phone lines but lacks the capacity of T1 lines. (See also *T1 line.*)

Linear pattern   Interactive pattern that encourages straightforward movement from screen to screen.

Management programs   Term used for programs that allow teachers to plan classes and track student development. Also used for similar programs used by administrative personnel in regard to employees.

Marinetti, F. T.   Prominent futurist and author of *The Futurist Manifesto* who first gave names to the concepts of simultaneity and multilinear lyricism.

Motion puzzle   A computer puzzle that involves rapid manipulation and placement of floating tiles or other moving pieces into a predetermined pattern.

MPEG   A compression scheme used to shrink video so that it can be stored on CD-ROM and played back later. (See also *QuickTime.*)

Multilinear lyricism   Futurist notion that saw divergent events, concepts, or objects as existing at the same time and in the same place in a radical juxtaposition made possible by advances in technology.

Multimedia   The combination of text, sound, and image in a computerized, interactive environment.

Narrative structure   Term used to describe the shape of a story and all of its constituent elements, including plot, character, setting, point of view, and theme.

Navigational tool   Tool or collection of tools used for the purpose of moving back and forth within an interactive program.

Needs analysis   Study of audience needs that occurs at the beginning of the instructional design process. (See also *Instructional design.*)

Obligatory cut   A screen or sequence of images placed in the interactive path in such a way that it must be viewed by users.

On-line education   Course work and supplemental educational material made available to students over a computer network.

Opening sequence   The opening screen, scene, or series of screens that comes at the beginning a program and sets its tone and purpose. Also known as a cut sequence.

Platform   Hardware system used to play, view, or manipulate a computer program.

Plot   The arrangement of events and characters' actions within a story.

Point of view   The angle at which events are perceived. Point of view can be that of an omniscient observer or can involve establishing perspective from within the frame of reference of a given character.

Process analysis   Form of classification in which information is categorized and presented according to a step-by-step model in which the various steps correspond to stages in a process.

Proposal   Initial document used to present basic conceptual ideas as a preliminary stage in winning approval or funding for a project.

Prototype   An early, typically simplified working model of a multimedia program that is created as production process begins. Prototypes are often text-only versions of a program that are created for alpha testing.

QuickTime   Video compression and playback software used on CD-ROMs and over the Internet. Frequent alternative to MPEG because it does not require special hardware on the playback computer.

Realm   A collection of screens or an extended screen.

Scene   A visual activity, usually dramatic in nature, that takes place on a given screen. A scene involves characters acting out an event in space and time. User may or may not have a role in the scene.

Screen   Fundamental design unit of interactive media. Movement on the computer is conceived as taking places between screens, each of which constitutes a separate entity or place.

Scripting process   The writing and design process from concept development to creation of a production script.

Setting   The place in which a program or part of a program occurs.

Simultaneity   Futurist concept concerning the juxtaposition of events, places, and people across temporal, physical, and cultural boundaries.

Spatial metaphor   Design metaphor in which the features, contours, and attributes of a place or an object are used as a way of conceptualizing the structure of an interactive program.

Strategy game   Computer game in which the primary task is making mental calculations regarding the position of objects in a spatial and temporal environment.

T1 line   High-bandwidth wiring system used to deliver multimedia from remote servers to workstations and personal computers.

Task   Something a user needs to accomplish on screen.

Theme   Meaning, intent, or ideative and emotional resonance associated with a given work.

Tools   Term used to describe the various objects or commands available to users for the manipulation of content or for navigation between content areas.

Treatment   An expanded concept document in which a writer or designer develops and presents ideas for content, interface, and interactive structure before writing the production script.

Twitch game   Computer shoot 'em up. (See also *Arcade game.*)

Web pattern   Interactive pattern in which a central screen is cross-linked to many other screens, many of which may be linked to each other.

Web site   A site or address on the World Wide Web. An individual site might consist of many interlinked screens (pages) and may be linked to other sites as well.

World Wide Web   Part of the Internet consisting of hyperlinked sites, each capable of supporting graphical user interfaces in conjunction with text, video, and audio.

Xanadu   Mythic realm, first envisioned by Theodor Nelson, in which all human knowledge is linked together in a vast, interconnected library that defies physical dimension.

SELECTED BIBLIOGRAPHY

Books and Periodicals

Benedikt, Michael, ed. *Cyberspace: First Steps* (Cambridge, MA: The MIT Press, 1992).

Benson, Thomas W., and Michael H. Prosser, eds. *Readings in Classical Rhetoric* (Bloomington: Indiana University Press, 1969).

Briggs, L. J. *Handbook for the Design of Instruction* (Pittsburgh: American Institutes of Research, 1970).

Brockman, John. *Digerati: Encounters with the Cyber Elite* (San Francisco: Hard-Wired, 1996).

Bush, Vannevar. "As We May Think," *Atlantic*, August 1945.

Cooper, Lane, ed. *The Rhetoric of Aristotle* (Englewood Cliffs, NJ: Prentice-Hall, 1932).

DeCecco, J. P. *The Psychology of Learning and Instruction: Educational Psychology* (Englewood Cliffs, NJ: Prentice-Hall, 1968).

Dick, Walter, and Lou Carey. *The Systematic Design of Instruction* (Glenview, IL: Scott, Foresman, 1985).

DiNucci, Darcy, ed. *The Macintosh Bible* (Berkeley, CA: Peachpit Press, 1994).

Donnelly, Daniel, ed. *In Your Face: The Best of Interactive Interface Design* (Rockport, MA: Rockport Publishers, 1996).

Farkas, Bart, and Christopher Breen. *The Macintosh Bible Guide to Games* (Berkeley, CA: Peachpit Press, 1996).

Field, Syd. *Screenplay: The Foundations of Screenwriting* (New York: Dell, 1979).

Franklin, Jon. *Writing for Story* (New York: Mentor, 1987).

Freeman, Judi. *The Dada & Surrealist Word Image* (Cambridge, MA: MIT Press, 1989).

Fromm, Ken, and Nathan Shedroff, eds. *Demystifying Multimedia* (San Francisco: Vivid Publishing, 1993).

Fuchs, Rainer, ed. *Self-Construction* (Vienna: Museum Moderner Kunst Stiftung Ludwig Wien, 1995).

Gagne, R. M. *Conditions of Learning* (New York: Holt, Rinehart & Winston, 1977).

Gagne, R. M., W. Wager, and A. Rojas. "Planning and Authoring Computer-Assisted Instruction Lessons," *Educational Technology* 21(9):17–26, 1981.

Gardner, John. *The Art of Fiction* (New York: Knopf, 1984).

Gibson, William. *Neuromancer* (New York: Berkeley Publishing Group, 1984).

Gleick, James. *Chaos: Making a New Science* (New York: Viking, 1987).

Janson, H. W. *History of Art* (New York: Harry N. Abrams and Prentice-Hall, 1973).

Kaufman, R., and F. W. English. *Needs Assessment: Concept and Application* (Englewood Cliffs, NJ: Educational Technology Publications, 1979).

Korolenko, Michael. *Writing for Multimedia* (Belmont, CA: Integrated Media Group, 1996).

Kreuger, Myron, *Artificial Reality* (Reading, MA: Addison-Wesley, 1983).

Kroll, Ed. *The Whole Internet: User's Guide & Catalog*, 2nd ed. (Sebastopol, CA: O'Reilly & Associates, 1994).

Laurel, Brenda, ed. *The Art of Human-Computer Interface Design* (Menlo Park, CA: Addison-Wesley, 1990).

———. *Computer as Theater* (Menlo Park, CA: Addison-Wesley, 1991).

May, Charles E. *Short Story Theories* (Athens: Ohio University Press, 1976).

McLuhan, Marshall. *Understanding Media: The Extensions of Man* (New York: McGraw-Hill, 1964).

Mok, Clement. *Graphis New Media 1* (New York: Graphis U.S., 1996).

Negroponte, Nicholas. *Being Digital* (New York: Random House, 1995).

Nelson, Theodor. "Interactive Systems and the Design of Virtual Reality," *Creative Computing*, November–December, 1980.

Norman, Donald A. *The Design of Everyday Things* (New York: Doubleday, 1988).

Payne, D. A. *The Assessment of Learning: Cognitive and Affective* (Lexington, MA: Health, 1974).

Perloff, Marjorie. *The Futurist Moment* (Chicago: University of Chicago Press, 1986).

Reiser, R. A., and R. M. Gagne. *Selecting Media for Instruction* (Englewood Cliffs, NJ: Educational Technology Publications, 1983).

Rheingold, Howard. *Virtual Reality* (New York: Touchstone Books, 1991).

Rivers, William L., and Alison R. Work. *Writing for the Media* (Mountain View, CA: Mayfield Publishing, 1988).

Russell, J. D. *Modular Instruction* (Minneapolis: Burgess Publishing, 1974).

Standsberry, Domenic. "Going Hybrid: The CD-ROM/On-line Connection," *New Media* (June 1995): 34–40.

Stevick, Philip, ed. *The Theory of the Novel* (New York: Free Press, 1967).

Stitch, Sidra. *Anxious Visions: Surrealist Art* (New York: Abbeville Press, 1990).

Sutherland, Ivan. "A Head-Mounted Three-Dimensional Display," *Proceedings of the Fall Joint Computer Conference* (1968).

Turkle, Sherry. *Life on the Screen: Identity in the Age of Internet* (New York: Simon & Schuster, 1995).

———. *The Second Self: Computers and the Human Spirit* (New York: Simon & Schuster, 1984).

Velthoven, Willem, and Jorinde Seijdel, eds. *Multimedia Graphics: The Best of Global Hyperdesign* (San Francisco: Chronicle Books, 1996).

Williams, Rubin. *Dada and Surrealist Art* (New York: Abrams, 1968).

CD-ROMs

EDUCATION

500 Nations, Mass Productions: Seattle.

1996 Encarta, Microsoft: Redmond, WA.

American Poetry: The Nineteenth Century, The Voyager Company: New York.

Amnesty Interactive, The Voyager Company: New York.

An Anecdotal Archive from the Cold War, George Legrady: San Francisco.

A Passion for Art, Corbis Publishing: Bellevue, WA.

Art Gallery: The Collection of the National Gallery, London, Microsoft: Redmond, WA.

Better Homes and Gardens' Complete Guide to Gardening, Multicom Publishing: Seattle.

Bio.morph Encyclopedia Muybridge, Nobuhiro Shibayama, 4D: Tokyo.

BodyWorks, Softkey International: Cambridge, MA.

Boxer Trigonometry, Boxer Inc.: Charlottesville, VA.

CNN Time Capsule, Vicarious: Redwood City, CA.

Cosmology of Kyoto, Softedge: Kobe, Japan.

The Day After Trinity, The Voyager Company: New York.

Dr. Seuss's ABC, Random House Children's Media: New York.

Events that Changed the World, ICE: Redwood Shores, CA.

The Foundation Marguerite and Aime Maiegt: A Stroll in Twentieth-Century Art, Matra Hachette Multimedia: Paris.

History of the World, Dorling Kindersley Multimedia: London.

House/Huis, Joost Grootens: Amsterdam.

Introduction to Multimedia: Multimedia University, Integrated Media Group: Belmont, CA.

Jumpstart Kindergarten, Knowledge Adventure: Glendale, CA.

Le Louvre: Palace and Paintings, Montparnasse Multimedia, Reunion des Musées Nationaux: Paris.

MacBeth, A. R. Braunmuller and David S. Rodes, The Voyager Company: New York.

Mapplethorpe—An Overview, Digital Collections, Inc.: Alameda, CA.

Marsbook CD, NASA: Houston.

Material World, Starpress Multimedia: San Francisco.

Math Munchers Deluxe, MECC: Minneapolis.

The Multimedia Cartoon Studio, Vanguard Media: New York.

Nile Passage to Egypt, Discovery Channel Multimedia: Bethesda, MD.

Ocean Life, Sumeria: San Francisco.

Oregon Trail II, MECC: Minneapolis.

Poetry in Motion, The Voyager Company: New York.

Redshift, Maris Multimedia: London.

Registros de Arquitectura 1: Mateo at ETH, Vicente Guallart, Producciones New Media SL: Barcelona.

Science Sleuths, MECC: Minneapolis, MN.

The Society of Mind, Marvin Minsky, The Voyager Company: New York.

Storybook Weaver Deluxe, MECC: Minneapolis.

Stowaway! Dorling Kindersley Multimedia: London.

Thinking Things, Edmark: Redmond, WA.

Troggle Trouble, MECC: Minneapolis.

The Ultimate Human Body, Dorling Kindersley Multimedia: London.

Vizability, PWS Publishing, Meta Design: San Francisco, CA.

The Way Things Work, David Macaulay, Dorling Kindersley Multimedia: London.

Women at Risk for HIV, Carolyn Sherins, Art Center College of Design: Pasadena, CA.

World Reference Atlas, Dorling Kindersley Multimedia: London.

The Wall Street Journal Review of Markets and Finance, Magnet Interactive, Washington, DC.

ENTERTAINMENT

The 7th Guest, Virgin Interactive: Irvine, CA.

Advanced Dungeons and Dragons, SSI/MacSoft: Sunnyvale, CA.

Afternoon, A Story, Michael Joyce, Eastgate Systems: Cambridge, MA.

Alice, Synergy Interactive: San Francisco.

All My Hummingbirds Have Alibis, Morton Subotnick, The Voyager Company: New York.

Amanda Stores, Amanda Goodenough, The Voyager Company: New York.

Arthur's Teacher Troubles, Marc Brown, Random House New Media: New York.

Bach and Before, Alan Rich, The Voyager Company: New York.

Bad Mojo, Vinny Carella, Pulse Entertainment: Santa Monica, CA.

Battlefront Series, Strategic Studies Group: Pensacola, FL.

Blender, Jenny von Feld: New York.

Civilization, MicroProse: Hunt Valley, MD.

Commander Blood, Microfolie's Editions: France, USA.

The Complete Maus, Art Spiegelman, The Voyager Company: New York.

The Daedalus Encounter, Virgin Interactive: Irvine, CA.

The Dark Eye, Edgar Allan Poe, Inscape: Los Angeles, CA.

Dark Forces, Lucas Arts: San Rafael, CA.

Dragon Lore, Mindscape: Novato, CA.

Dreggs, John A Kuri, Fathom Studios: Sausalito, CA.

Eight Ball Deluxe, Little Wing, Amtex Corporation: Belleville, Ontario, Canada.

Elroy Goes Bugzerk, headbone interactive, Inc.: Seattle.

Firefall Arcade, Focus Enhancements: Woburn, MA.

Flashback, MacPlay: Irvine, CA.

Flying Nightmares, Domark: San Mateo, CA.

Full Throttle, Lucas Arts: San Rafael, CA.

Gadget, Synergy Interactive Corporation: San Francisco.

A Hard Day's Night, The Beatles, The Voyager Company: New York.

Indiana Jones and the Fate of Atlantis, Hal Barwood, Lucas Arts: San Rafael, CA.

In the 1st Degree, Peter Adair, Haney Armstrong, and Domenic Stansberry, Broderbund: Novato, CA.

I Photograph to Remember/Fotografio Para Recordar, Pedro Meyer, The Voyager Company: New York.

Just Grandma & Me, Mercer Mayer, Random House Children's Media: New York.

Madness of Roland, Greg Roach, Hyperbole Studios: Seattle.

Media Band, Canter Technology: San Francisco.

Monopoly, MacPlay: Irvine, CA.

Myst, Broderbund: Novato, CA.

Passage to Vietnam, Rick Smolan, Against All Odds Productions, Sausalito, CA.

Playmaker Football, Playmaker, Inc.: Phoenix, MD.

Power Poker, Electronic Arts: San Mateo, CA.

Prince of Persia, Broderbund: Novato, CA.

The Puppet Motel, Laurie Anderson with Hsien Chien Huang, The Voyager Company: New York.

Putt Putt, Annie Fox, Humungous Entertainment: Woodinville, WA.

Rebel Assault, Lucas Arts: San Rafael, CA.

The Residents, The Voyager Company: New York.

Richard Strauss: Three Tone Poems, Russell Steinberg, The Voyager Company: New York.

The Riddle of the Maze, Fathom Pictures: Sausalito, CA.

Scrutiny in the Great Round, Calliope Media: Santa Monica, CA.

SimCity, Will Wright, Maxis: Walnut Creek, CA.

Strategic Conquest, Delta Tao Software: Sunnyvale, CA.

Super Wing Commander, Origin Systems: Austin, TX.

Tetris, Spectrum Holobyte: Alameda, CA.

Victory Garden, Stuart Moulthrop, Eastgate Systems: Cambridge, MA.

Warlords II, Strategic Studies Group: Pensacola, FL.

Who Killed Sam Rupert? Shannon Gilligan, Creative Multimedia Corporation: Portland, OR.

Xplora1: Peter Gabriel's Secret World, Brilliant Media: San Francisco.

Zone Warrior, Cassady & Greene: Salinas, CA.

The Zork Anthology, Activision: Los Angeles.

Zork Nemesis, Activision: Los Angeles.

BUSINESS AND PROMOTIONAL

2Market, 2Market, Inc.: Chatsworth, CA.

The 1994 IBM Interactive Annual Report, International Business Machines Corporation: Atlanta.

Big Hand Digital Portfolio, Big Hand Productions: Dallas.

From This Point Forward, NYNEX, Magnet Interactive Studios, Washington, DC.

Haas School of Business: Interactive Tour, Haas School of Business, University of California: Berkeley.

imedia Studio Tour, Brian Almeshie, 3D JOE: San Francisco. (Floppy disk)

Informer One, Colin Taylor and Jonathan Taylor, .nformer Interactive Research: London.

I/O 360 Demo Thing, i/o 360 Digital Design: New York.

The Leonhardt Group, The Leonhardt Group: Seattle.

May & Company Portfolio, May & Co.: Dallas.

Mercedes-Benz Interactive, Mercedes-Benz of North America, The Desinory, Inc.: Long Beach, CA.

Multimedia Business 500, Allegro New Media: Fairfield, NJ.

Tanagram Digital Postcard, Tanagram: Chicago.

Toyota Interactive, Toyota Motor Sales, USA, Saatchi & Saatchi DFS/Pacrici: Torrance, CA.

Toyota Tacoma Launch Ad, Toyota Motor Sales, USA, Saatchi & Saatchi DFS/Pacrici: Torrance, CA.

Welcome to CompuServe, CompuServe: Columbus, OH.

ZeitMovie & Selected Notes Zeitguys, Bob Aufuldish, Zed: The Journal of the Center for Design Studies at Virginia Commonwealth University: Norfolk, VA.

Zip Tour, Iomega Corporation, Fitch Inc: Worthington, OH.

Web Sites*

EDUCATION

American Memory Website, Library of Congress. http://rs6.loc.gov/amhome.html

American Red Cross. http://redcross.org

Contemporary Realistic Gallery Website. http//www.realart.com

Cyber Film School. http://www.cyberfilmschool.com

Dade County Public Schools Website. http://dcps://dade.k12.fl.us/www/inst/main.html

EnviroLink. http://www.envirolink.com

Exploratorium. http://www.exploratorium.edu

Facets of Religion. http://www.christusrex.org

Frog Dissection Activity, University of Virginia, Department of Technology. http://curry.edschool.Virginia.EDU/go/frog/

GINA: Global Initiative for Asthma. http://www.ginasthma.com/asthma/GINANEW.html

Gulf War Veteran Resource Pages. http://www.gulfwar.org/index.html

Gulf War Veterans. http://www.ed.gov/prog_info/SFA/StudentGuide

Health World Online. http://www.healthy.net

History of Home Video Games Homepage. http://www.sponsor.net/~gchance/

The Motley Fool. http://fool.web.aol.com/fool_mn.html

OneWorld Online. http://www.oneworld.org

Origins of Humankind. http://www.dealsonline.com/origins

Politics Now, ABC News, et al. http://www.politicsnow.com

The Shakespeare Homepage, Massachusetts Institute of Technology. http://the-tech.mit.edu/Shakespeare/index.html

Student Guide, The World Wide Web Virtual Library: Museum. http://www.comlab.ox.ac.uk/archive/other.museums.html

*Because Web site addresses and locations change frequently and without notice, the author cannot guarantee the ongoing viability of all Internet addresses listed here.

The Teel Family WWW Site. http://www.teelfamily.com

UC Museum of Paleontology, University of California, Berkeley. http://www.ucmp.berkeley.edu

The Voice of the Shuttle, University of California, Santa Barbara. http://humanitas.ucsb.edu

Youth Central. http://www.youthcentral.apple.com

Yu Hsin's Den. http://merlion.signet.com.sg/~seahyh/

ENTERTAINMENT

Addicted to Noise. http://www.addict.com

Berit's Best Sites for Children. http://www.cochran.com/theosite/Ksites.html

The Bingo Zone. http://www.bingozone.com

Catapult/X-Band. http://www.exband.com

Chessmaster 5000. http://chessmasternetwork.com

CineMania Online. http//204.255.247.126:80/cinemania

The Classic Typewriter Page. http://xavier.xu.edu:8000/~polt/typewriters.html

The Comic Strip. http://unitedmedia.com/comics

*Cyberia*2, Prima Publishing. http://www.fwa.com/index.html

The Digital City, Stichting De Digital Stad. http://www.dds.nl

Electra Studios Presents: Suspect! http://www.electrastudios.com/suspect

Engage Games Online. http://www.gamesonline.com

Epicurious. http://www.Epicurious.com

ESPNet SportZone. http://espnet.sportszone.com

Fine Line Features. http://www/flf.com

Fox.com, 20th Century Fox. http://www.fox.com

Gamesmania. http://www.gamesmania.com

Gamespot. http://www.gamespot.com

HomeArts, The Hearst Corporation. http://homearts.com/depts/fresh/hewhome.html

HotWired. http://www.hotwired.com

Interplay Productions. http://www.vie.com

Madeline's Mind. http://www.madmind.com

Mplayer. http://www.mplayer.com

My Boss. http://www.myboss.com

National Park Service. http://www.nps.gov

North American Steam Locomotives. http://www.arc.umn.edu/~wes/steam.html

Official Unofficial Nash Bridges Page. http://www/lowtek.

President '96. http://www.pres96.com

Rancho DeNada. http://users.accesscomm.net/~rancho/index.html

The Spot. http://www.thespot.com/

Storyweb. http://www.storyweb.com

The Strange Case of the Lost Elvis Diaries. http://home.mem.net/~welk/elvisdiaries.html

Suck. http://www.suck.com

Total Entertainment Network. http://www.ten.net
Trilobyte. http://www.tbyte.com
Ultima Online. http://www.owo.com
Videomaker Camcorder and Desktop Video Site. http://www.videomaker.com
Women's Wire Website. http://www.women.com
The X-Files. http://www.thex-files.com

BUSINESS AND PROMOTIONAL

The Advertising Age Website. http://www.adage.com
Apple Computer. http://www/apple.com./
Arthur Andersen Business Consulting Website. http://www.arthurandersen.com/new home.html
Atomic Vision. http://www.atomicvision.com/rotary.html
Business Resource Center. http://www.kciLink.com/brc/
Career Mosaic. http://www.careermosaic.com
Car Talk. http://cartalk.com.
C/Net World Wide Website. http://wwwcnet.com
Creative Zone. http://www.creat.com/zonemenu.html
David Byrne Website. http://www.bart.nl/~francey/th_db2.html
Delta Airlines Website. http://www.Delta-air.com
Discovery Channel Online. http://www.discovery.com
Electra Entertainment Website. http://www.elektra.com
Epson America Website. http://www.epson.com
Evergreen Network. http://cybermart.com
Home Financial Network Website. http://www.movado.com
Idea Cafe: The Small Business Channel. http://www.ideacafe.com
Intel. http://www.intel.com//
ISDM Website. http://www.ping.be/Zupergraphyx!
JobSmart. http://jobsmart.org/
Kellogg Website. http://www.kelloggs.com
Macromedia Website. http://www.macromedia.com
Photodisc World Wide Website. http://www.photodisc.com
Primo Angeli Inc. Website. http://www.primo.com
The Saturn Site. http://www.saturncars.com.
Shannon Roach's Personal Website. http://www.disonline.com/userweb/shannonr
Silicon Surf. http://www.sgi.com
Spiegel Directions. http://www.spiegel.com
The Warren Idea Exchange. http://www.warren-idea-exchange.com

INDEX

Adobe *Illustrator*, 167
Adventure, 97, 98
Adventure games, 90, 96–98
Adventure Simulator, 168
Advertising programs
    audience and, 136
    guide figures for, 139
    obligatory cuts in, 140
    opening screens for, 137–138
    purpose of, 136, 140
    shopping brochures or
        catalogues as, 136, 139
    structure considerations,
        136–137
    text vs. image, 139–140
Aesthetics
    futurist movement and, 3–7
    multilinear lyricism and, 4
    simultaneity and, 3–4
    study in multimedia field, 12–13
Agrippa (Gibson), 26
Alice-in-Wonderland avatars,
    99–100
Alpha test, 21
America Online, 139
American Memory Web Site,
    42, 43
Analysis as teaching method,
    118–119
Anatomy lab, on-line, 130, 131
Anderson, Laurie, 74, 75
Antagonists, 73
Apollinaire, Guillaume, 3, 172
Apple Computer, 39
ARC (Augmentation Research
    Center), 3, 9
Arcade (*twitch*) games
    character metaphor in, 39
    computer's impact on, 93
    critical discussions of, 78
    in educational multimedia, 122

event metaphor in, 38
historical roots of, 92–93
maze games, 62, 93–94
on-line, 98
player's expectations in, 101
skills required for, 90
Aristotle
    on classification, 41
    on narrative structure, 11
    on plot and character, 72, 73
The Art of Fiction (Gardner), 33
Arts and sciences
    challenges to Newtonian
        views in, 7, 8–10
    computers and, 3–10
    dadaist movement in, 4–5
    futurist movement in, 3–7
    surrealist movement in, 4–5
As I Lay Dying (Faulkner), 77
"As We May Think," 9
Asimov, Isaac, 170
Assessment, 119, 120, 133
Associative logic, 71–72
Asteroids, 93
Atomic Vision, 138
Audience
    of advertising multimedia, 136
    of computer games, 101–102
    consumers as, 136–137
    of educational multimedia,
        113–115, 133
    new roles for, 25–26
    research, 20–21
    testing, 21, 27–28, 30, 128
Audio
    computer memory and, 165
    format capabilities for, 163–164
    on the Internet, 100, 158,
        161–162
Augmentation Research Center
    (ARC), 3, 9

Authoring programs, 167–168
Authorware, 167
Avatars, 99–100

Background material (narration),
    81–82
Bandwidth, 98–99, 150, 161–162,
    163
Beneficial Finance, 145
Beta test, 21
Bible, character, 103–105
Big Brother, 173
Blue-screen process, 98, 160–161
Board games, 90, 91, 94
Boccioni, Umberto, 172
Books as design metaphor, 37
Bradbury, Ray, 170
Branching structures
    broad and shallow, 64
    crisscrossing, 66, 67
    deep, 64, 65
    evaporating, 66, 67
    forking, 66, 67
    multiple, 65–67
    stepladder, 68
Braunmuller, A. R., 160
Brave New World (Huxley), 170
Bridge (card game), 94
Broad and shallow branching
    structure, 64
Brochures
    flowchart example for, 45
    shopping, 136, 139
Broderbund, 23, 39, 106, 121, 161
Brothers, 40
Bruner, Jerome, 10, 12, 13, 53
Bush, Vannevar, 9–10, 12
Business programs
    advertising programs, 136–140
    design process for, 150–153

flowchart example for, 142–143
format considerations in, 148–150
informational programs, 140–145
Intranet systems, 150
production scripts for, 153
proposal stage in, 150–151
screen-by-screen treatment for, 151–153
software for writers/designers of, 167–169
training programs, 145–148
unique characteristics of, 135–136
writer's role in, 154
Byrne, David, 22

C programming language, 167, 169
Cable modems, 162
Canasta (card game), 94
Canter Technology, 36
Card games, 90, 94
Catalogues
floppy disk vs. CD-ROM, 149
shopping, 136, 139
CD-ROM format
floppy disk vs., 149
for interactive media, 18, 19
software for writing in, 167, 169
technical aspects, 157–159, 162
video display on, 160–161, 164
Cendrars, Blaise, 3, 172
Chaos theory, 8
Character bible, 103
Characters
in adventure games, 96, 98
Aristotle on, 72, 73
defined, 73
as design shell orientation, 39–40
detailed descriptions of, 103
developing, in interactive mediums, 75–76
lists, example of, 104–105
users as, 76–77
Charting. *See also* Flowcharts
multilinear dialogue, 108–110
software for, 166–169
Chat rooms, 99, 100, 130, 131
Checkers (game), 90, 94

Chess (game), 90, 94
Chomsky, Noam, 173
Ci-gît Breton (Desnos), 5
Cicero, 41, 79
Cinema, interactive. *See* Interactive cinema
Classification schemes
defined, 22, 41
as design shell, 40–44
drawbacks of, 43
general rules about, 43
in informational programs, 144
screen examples of, 22, 42, 43, 44, 149
sorting by attributes, 41–42
sorting by topic or type, 42
CNN on-line video clips, 130
Collage art, 4, 5, 75, 76, 171, 172
Collins, Wilkie, 103
Comparison/contrast, 42, 82–84
Competitive element in games, 90–91
Compression schemes, 159–161, 174
Computer Curriculum Company, 121
Computer games. *See also* Games
arcade, 38, 78, 92–94, 101, 122
audience and, 101–102
competitive element in, 90–91
concept development for, 100–102
concept premise for, 101–102
concept treatment for, 102–107
historical background of, 88–92
industry characteristics, 89
maze games, 62, 93–94
motion puzzles, 94, 95
on-line gaming, 98–100
physical space renderings in, 91, 92
premise statement for, 101–102
production script example for, 107–108
scripting process for, 100–110
software for writers/designers of, 168
strategy games, 94–96
text-based, 97, 98
types of, 92–100
war games, 95
Computer memory, 164–165

Computer sciences, early, 9–10
Computers. *See also* Technical considerations
in classrooms, 132–133
differentiated from other media forms, 2
first personal, 10, 12
Futurist vision of, 7, 10
as a medium, 2
Concept development
for computer games, 100–107
definition stage in, 18–20
design process and, 17, 18
design shell stage in, 21–23
in instructional design, 115–117
research stage in, 20–21
technical literacy and, 156–157
Concept premise, defined, 101–102
Concept statement, defined, 20
Concept treatment. *See* Treatment
Concluding elements, in expository structure, 84–85
Conditional linking, 130, 165
Consumers as audience, 136–137
Content
defined, 33
documents relating to, 23–25, 26–27
form vs., 32–34
in instructional design, 118–121
interface design and, 47–48
Content arrangement
classical notions of, 71–72
comparison/contrast, 83–84
expository, 79–85
hierarchical, 74–75
interactivity's impact on, 71–72
linear, 73–74
main body of, 82–84
metaphors and, 34–40
narrative, 72–79
plot and character in, 72–76
point of view and, 76–77
process analysis in, 82–83
rhetorical techniques for, 82–84
setting, style, and theme in, 77–78
value of classical techniques of, 72, 85
web-based patterns of, 74–75
Content experts, 114, 118, 128, 132

Content outline, defined, 114–115
Content research, 20–21
Content treatment. *See* Treatment
Conversations, between user and character, 76
Corporate training. *See also* Business programs
  elements in, 145–148
  flowchart example for, 147
Course outline development, 119–120
Courtroom programs. *See also In the 1st Degree*
  as event-oriented metaphor, 38–39
Crawford, Chris, 168
Creative personnel, various job titles for, 16
Crisscross branch structure, 66, 67
Cross-linking, 62
Crystal Caliburn, 92
Curricula design for the Internet, 129–132
Cursor
  dexterity skills and, 101
  as game implement, 92, 93

Dadaist movement, 4–5, 172
Dade County Public Schools Web Site, 42
The Daedalus Encounter, 98
Decision points, 55–56, 69
Delta Airlines Web Site, 137, 138
Demonstration versions, 28, 128
Depth vs. breadth in interactive structures, 64, 65
Der Dada #3 (Heartfield), 6
Design by symmetry, 35
Design documents
  defined, 27, 124
  types of, 20, 23, 27, 103, 115
Design metaphor
  for advertising programs, 136
  book as, 37–38
  classification scheme vs., 40–44
  defined, 22, 35
  design shell and, 22, 34–40
  events as, 38–39
  in instructional design, 121–123
  limitations of, 37–38
  maps as, 37

objects as, 37–38
screen examples of, 23, 36, 38, 39, 48
spatial, 36–38
two-dimensional characters in, 39–40
Design process
  for business programs, 150–153
  for computer games, 100–110
  concept development stage in, 18–23
  elements of, 16–17, 29
  overlapping roles in, 16–17
  overview of, 15, 17–18
  production stage, 27–28
  script design stage, 23–27
  skills used in, 16
  writer's role in, 15–18, 29–30
Design shell
  classification scheme as, 40–44
  concept development and, 21–23
  as conceptual tool, 21–23
  conducting research on, 20–21
  defined, 21, 34, 49
  dual design schemes, 43–44
  elements of, 21
  end goal of, 35
  familiar activities as, 121–122
  flowcharts as tool for, 44–46
  form vs. content in, 32–34
  immersive environments as, 122, 124
  in instructional design, 121–123
  interface considerations in, 46–49
  metaphors as, 22, 34–40
  practical aspects of, 48–49
  process and tools as, 122–123
  as stage in design process, 18
  technological limitations of, 49
Design team
  overlapping roles in, 17
  personnel of, 16, 28–29
Designer's Edge, 167–168
Desnos, Robert, 4, 5
Dialogue. *See also* Script design; Scripting process
  flowcharting multilinear, 108–110
  scripts, 27
Digerati, 180

Digital Revolution, 3, 176–177
Digital video disk (DVD), 18, 19, 159, 164
Division of contents, in expository structure, 80–81
Drama. *See also* Interactive cinema
  elements of, 72–76, 85
  on-line soap operas, 99–100
Dungeons & Dragons, 90
DVD (digital video disk), 18, 19, 159, 164
Dynabook, 10, 12

Eastgate Systems, 168
Educational multimedia. *See also* Instructional design
  defined, 112
  future of, 132–133, 177–179
  the Internet and, 129–132
  software for writers/designers of, 166–169
  types of, 112–113
  writer/designer's role in, 132–133
Eiffel Tower, 4
Einstein, Albert, 8
Electronic Arts, 101
Embellishment of screens, 56, 153
Energeia, 73
Engelbart, Douglas, 3, 9, 12
Entertainment multimedia
  creative personnel in, 16
  software for, 166–167
Erickson, Thomas D., 35
Ernst, Max, 172
Evaluation techniques, 128–129
Evaporating branch structure, 66, 67
Event-oriented metaphors, 38–39
Experts, content, 114, 118, 128, 132
Exploratory programs
  defined, 112–113
  example of, 145
  interface style, 78, 144
Expository structure
  background material in, 81–82
  concluding elements in, 84–85
  division of contents in, 80–81
  interactivity's impact on, 25–26

introductory material in, 79–81
main body in, 82–84

Fascism, 172–173
Faulkner, William, 77
Fiber-optic lines, 130, 174
Fiction, software for writing
    interactive, 168
Fictional personas, 99–100
Field trials, 129
Flashback, 101, 102
Flight simulator training
    programs, 38, 122
Floor plans, as design metaphor,
    37
Floppy disk format
    CD-ROM vs., 149
    characteristics of, 18, 19
    technical aspects, 157–159
Flowcharts
    as conceptual tool, 63, 69
    in design shell development,
        44–46
    in early stages of script
        design, 23
    example of simple, 45
    for informational programs,
        142–143
    limitations of, 66, 68
    purpose of, 44
    software for creating,
        166–167, 169
    for task-oriented training
        programs, 147–148
Focus groups, 129
Ford, Henry, 172–173
Forking branch structure, 66, 67
Form
    content vs., 32–34
    defined, 33
    interface design and, 47–48
Format. *See also* Technical
    considerations
    audio and, 163–164
    choices available, 149, 157–159
    defined, 157
    major types of, 18, 19
    video and, 160–161, 164
Foundational programs, 112–113
Fractals, 8–9

Free association, navigation
    process and, 71–72
Frog Dissection Activity, 130, 131
Fujitsu, 99
Full treatment, example of, 107
Functionality, 27, 28, 153
The Futurist Manifesto
    (Marinetti), 3–7, 12, 171
Futurists
    history of, 3–7
    interactive multimedia and, 179
    symbol of, 4
    views on technology, 171–173

Game play, in treatment, 103, 105
Gamers, defined, 89
Games. *See also* Computer games
    common elements in, 90–92
    competition in, 90–91
    defined space or territory in, 91
    historical background of, 88–89
    inventory vs. implement
        accumulation in, 91
    nature of, 89–92
    psychological aspects of, 88–89
    rules and, 91
    skills required for, 90
    treatment example for, 102–107
Gardner, John, 33
Georgia, 74
Gibson, William, 26, 170, 180
Gleick, James, 8
Glossaries, 66, 82, 83
Graphics
    design and, 16, 29
    on the Internet, 163–164
Guide figure
    in advertising programs, 139
    as design metaphor, 39–40
Guided tours, 141, 144
"Guides project," 39

Haas School of Business
    interactive tour, 141, 144
Hands-on learning, 118
Hardwired, defined, 130, 165
Harrington, Michael, 173
Heartfield, John, 4, 6
Hearts (card game), 94

Hierarchical patterns
    characteristics of, 58–59
    defined, 57
    flowchart example of, 60
Home base, 47
Hot spots
    defined, 56
    in game scripting, 108–109
    hyperlinking of, 61–62
    in linear patterns, 58
    in web-like patterns, 61–62
HTML (Hypertext Mark-up
    Language), 167
Huxley, Aldous, 170
HyperCard, 167, 169
Hyperlinking
    defined, 55–56
    in open-ended web patterns,
        61–62
    to glossaries, 66, 82
HyperStudio, 167
Hypertext mark-up language
    (HTML), 167

I Photograph to Remember, 73, 74
IBM PCs, 157
Icons, defined, 56
Immersive environment
    in instructional design, 122, 124
    in job-training programs,
        145–148
    as teaching method, 118–119,
        124
Implementation of educational
    multimedia, 128–129
In the 1st Degree, 19, 39, 40, 74,
    98, 106, 157, 161
Indexing systems, in scripting
    process, 109–110
Inform, 168
Information designers, 29
Informational catalogues, 149
Informational programs
    flowchart example for, 142–143
    obligatory cuts in, 141
    opening screens for, 141, 144
    purpose of, 140–141
    structure considerations,
        142–143
Inspiration, 166

Installation art, 2, 4, 6, 8
Institutional interactive
    programs, 140–145
Instructional design
    analytical vs. immersive
        environments for, 118–119
    audience profile in, 114–115,
        133
    basic elements of, 113, 133
    content development stage,
        117–121
    content document for, 119–120
    content outline, 114–115
    course outline development
        in, 119–120
    design metaphors in, 121–123
    evaluation and testing in, 119,
        128–129
    flowchart for task-oriented
        training program, 147–148
    implementation stage in,
        128–129
    interactive structure (design
        shell) for, 121–123
    mission statement, 115–117
    modules in, 119–120
    motivation of students in, 115
    needs analysis, 113–115, 133
    production scripts for, 126–128
    proposal stage in, 113–117
    revision stage in, 128–129
    skill sets, 118–119
    software for, 167–169
    teaching methods in, 118–119
    treatment writing in, 123–126
Instructional goals
    defined, 113, 115, 116
    example of, 116–117
Interactive brochures, 45, 136, 139
Interactive cinema
    defined, 38
    event-oriented metaphors in,
        38–39
    examples of, 39, 98
    on-line dramas, 99–100
    scripting and, 103
    soap operas, 99–100
    storytelling/character
        development in, 98
Interactive conceit, 20
Interactive designers, 29

Interactive media. *See also*
        specific types
    on computer vs. TV screens,
        175–176
    conflicting views on, 170–173,
        182
    critical reviews of, 78
    future directions for, 173–174
    long-range effects of, 170–182
    major formats for, 18, 19
    social costs of, 173–174
    technological advances and,
        174–176
    writer/designer role in, 15–17,
        182
Interactive structure. *See also*
Design shell
    adding complexity to, 63–69
    branching patterns of, 64–68
    classification schemes in, 22
    computer memory and,
        164–165
    conditional linking, 130, 165
    content/concept statements
        and, 20
    defined, 18, 24, 56
    design metaphors in, 22, 23
    embellishment of screens, 56,
        153
    hierarchical patterns of, 58–60
    hyperlinking in, 55–56, 61–62
    in instructional design, 121–123
    linear patterns of, 57–58
    navigating between screens,
        55–57
    pathways through, 25–26,
        46–47
    patterns of, 56–63
    screens and, 54–57, 69
    as treatment element, 24, 25,
        103, 105
    web-like patterns of, 61–63
*Interactive Tour of the Haas
    School of Business*, 141, 144
Interactivity. *See also* Interactive
        structure
    basic elements of, 54–56
    defined, 2, 12, 52–53
    flowcharting, 56, 63, 66, 68,
        106–107
    levels of, 25–26, 52–53

    patterns of, 56–63
    psychological aspects of,
        25–26, 54, 69
    structural units of, 54–56
    user control and, 53–54
    writer's role in, 11–12
Interface
    defined, 16, 21–22, 46
    design shell vs., 46–47
    as element in treatment, 24, 25
    styles of, 78
Interface design. *See also* Design
        process
    defined, 24, 46–48
    in game treatment, 103, 105
    primary concerns of, 47–49
    using flowcharts in, 106–107
Internet. *See also* World Wide Web
        (www)
    bandwidth and, 98–99, 150,
        161–162, 163
    business programs on, 149
    educational curricula on,
        129–132
    as interactive media format,
        18, 19
    on-line gaming and, 98
    as a presentation format, 158
    video and sound on, 130,
        163–164
Intranet systems, 150
Introducing Media Band, 36
Introductory elements, 79–81
Intuitive interface style, 78
Invention, in interactive media,
        17, 20
ISDN lines, 130, 162

Java, 98, 161
The Journey Man Project, 98
Joysticks, 93
Jumpstart First Grade, 38, 122,
        123
Juxtaposition in art, 4, 7, 172
    examples of, 5, 6, 7, 75

Kay, Alan, 10, 12, 36
Keyboards, as game implement, 93
Kick-off meetings, 150

Knowledge Adventure World, 38, 99, 122, 123
Kurosawa, Akira, 77

La Place, Pierre, 9
Le Guin, Ursula, 170
Library of Congress American Memory Web Site, 42, 43
Linear patterns
  characteristics of, 57–58
  flowchart examples of, 58, 59
  modified, 58, 59
Linking, 55–57
  conditional, 130, 165
  cross-linking, 62
  hyperlinking, 61–62
Literary critics, on impact of computers, 3
Literature
  futurist movement in, 5–6
  narrative structure in, computers and, 11–12
Logic, associative, 71–72
Lucas Film Habitat project, 99

Macbeth, 79, 80, 81, 82, 160
Macintosh computers, 22, 43, 157
McLuhan, Marshall, 10, 12–13, 33
Macromedia Director, 167, 169
Madeline's Mind, 100
Main body, in expository structure, 82–84
Main menu, defined, 47
The Making of In the 1st Degree, 161
Management programs, 112–113
Mandatory viewing screens. *See* Obligatory cut
Mandelbrot, Benoit, 8–9
Maps, as design metaphor, 37
Marinetti, Filippo Tommaso. *See also* Futurist movement
  on simultaneity/multilinear lyricism, 3–4, 12
  on technology, 171–172
Market research, 21
Marketplace, Internet as distribution center for, 177

Massachusetts Institute of Technology (MIT), 9
The Massacre of the Innocents (Ernst), 172
Maxis, 96
Maze games, 62, 93–94
MECC, 44, 124
MECC 1996 Sampler, 44
Media elements, 24, 25, 103, 105
Memory, computer, 164–165
Menu choices, defined, 56
Metaphors. *See also* Design metaphor
  defined, 35
  event-oriented, 38–39
  limitations of, 37–38
  spatial, 36–38
Meyer, Pedro, 73, 74
Minsky, Marvin, 40
Mission statement
  defined, 115
  example of, 116
  instructional goals in, 116–117
  topics covered in, 117
MIT (Massachusetts Institute of Technology), 9
Modules
  defined, 119
  example of outline for, 120
Monopoly (board game), 90, 91, 94
The Moonstone
  charting dialogue example, 109–110
  preliminary treatment example, 104–106
  production script example, 107–108
Motion puzzles, 94, 95
Motivation
  of characters, 75
  in instructional design, 115
Mouse, as interactive cognitive tool, 36
MPEG compression, 159–160
Multilinear lyricism
  artists and, 12, 171, 179
  in contemporary science, 9
  defined, 3–4
Multimedia. *See also* Interactivity
  aesthetic of, 1–14

business. *See* Business programs
  defined, 2, 12, 15
  educational. *See* Educational multimedia
  first associated with computers, 2
  future of, 176–182
  guiding principles of, 1–2
  nature of, 10–11
  as a new discipline, 1–2
  origins in arts and sciences, 2–10
  origins of, 3–10
  social aspects of, 170–176
  writer's role in, 11–12
Multiple and parallel branching structure, 65–66
Muncher games, 122
Mussolini, Benito, 172–173
Myst, 23, 37, 77, 98

1984 (Orwell), 170
Narration (background material), 81–82
Narrative structure
  Aristotle on, 11
  characters in, 72–76
  as content arrangement, 72–79
  elements in, 72–79
  Futurists on, 11–12
  interactivity and, 25–26
  for multimedia, 11–12
  plot in, 72
  point of view in, 76–77
  setting in, 77
  style in, 77–78
  theme in, 78
Navigation-heavy structure, 59
Navigational tools
  associative logic and, 71–72
  defined, 47
  fundamental choices of, 54–55, 69
  in game treatment, 103, 105, 106
  interface design and, 47–48, 55–56
Needs analysis, 113–115, 133

Nelson, Theodor, 10
Neuromancer (Gibson), 170, 180
Newtonian physics, impact of
    computers on, 7–10

Object-oriented programming, 36
Obligatory cut
    in computer games, 107–108
    defined, 64–65
    in informational programs, 141
Observation of Observation:
    Uncertainty (Weibel), 8
On-line environment. *See also*
    Internet
    dramas in, 99–100
    for educational curricula,
      129–132
    format capabilities and, 18, 19,
      149–150, 163–164
    gaming, 98–100
    scripting and, 103
    transmission limitations on,
      161–162
On-the-job training programs,
    145–148
Opening screens
    in expository programs, 79–80
    as first impression, 137–138,
      149
    in informational programs,
      141, 144
    as obligatory cut, 64
    to establish corporate image,
      149
Oregon Trail, 122, 124
Orwell, George, 170, 173

PacMan, 93–94, 122
Page toggles, 58
Paik, Nam June, 6
Palo Alto Research Center, 10
Parade Amoureuse (Picabia), 4, 7
Parallel branching structure,
    65–67
Parlor games, 90, 94
Perloff, Marjorie, 3
Photoshop, 167
Physical space

as design metaphor, 36–38
renderings in computer
    games, 91, 92
Physics, Newtonian, computer's
    impact on, 7–10
Piaget, Jean, 10, 13, 53
Picabia, Francis, 4, 7
Picasso, Pablo, 77
Place, as design metaphor, 37
Platforms, 157–159
Playback software, 159–160
Plot, in narrative structure,
    72–73
Point of view, in narrative
    structure, 76–77
Pound, Ezra, 3, 172
Preliminary treatment
    defined, 103
    example of, 104–106
Premise statement, 101–102
Process analysis, 42, 82–83
Production
    prototypes in, 28, 128
    as stage in design process, 17
    user testing and, 27–28
Production scripts
    for business programs, 153
    examples of, 107–108, 127–128
    format capabilities and, 164
    in instructional design,
      126–128
    technical aspects, 164
Program content. *See also*
    Treatment
    defined, 24–25
Programmers
    roles of, 29, 30, 154, 163
    writer's technical literacy and,
      156–157
Programming languages, for
    multimedia writers, 167, 169
Proposals
    in business programs, 150–151
    in instructional design, 113–117
Protagonists, 73
Prototypes, 28, 128
    authoring programs for,
      167, 169
The Psychic Detective, 98
Psychological components

of advertising programs,
    137, 140
of computer vs. TV screens,
    175–176
of games, 88–89
of interactivity, 25–26, 54, 69
of settings, 77–78
Puppet Motel, 74, 75

Quantum theory, 8
QuickTime, 160–161, 165

Rashomon (Kurosawa), 77
RealAudio, 158, 161–162
Realm, defined, 57
Relativism, 8, 179–181
Research
    audience, 20–21
    design shell, 18, 20–21
    market, 21
Return to Zork, 97–98
Revision, in instructional design,
    128–129
Rheingold, Howard, 9
Rhetorica Ad Herennium
    (Cicero), 41, 79
Rhetorical techniques, 82–84
    classification, 82
    comparison/contrast, 83–84
    definition (glossaries), 83
    example, 83
    process analysis, 82–83
Risk (board game), 94
Rodes, David S., 160
Role-playing games, 90, 96–98
    in educational multimedia, 122
Rubik's Cube, 94

Sales concept, 137, 140
Scene
    defined, 57
    scripts for, 27
Scholastic, 121
Science Sleuths, 48
Sciences. *See* Arts and sciences
Screens
    defined, 54

embellishment vs. hyperlinking, 56, 153
graphic materials and size of, 164
main, 57, 138
navigational choices and, 54–56, 69
obligatory cut, 64–65, 107–108, 141
opening. *See* Opening screens
psychological aspects of, 175
realms vs., 57
scenes vs., 57
Script design
content treatment and, 23–25
defined, 17
early documents in, 23
flowcharting dialogue in, 108–110
indexing systems for, 109–110
interactivity and, 25–28
nonsequential elements in, 108–110
production scripts, 26–27, 30, 107–108
types of scripts, 26–27
Script Thing, 166
Scripting process. *See also* Treatment
for advertising programs, 136–140
basic elements of, 102–103, 110
for business multimedia, 135–136
for computer games, 100–110
for corporate training programs, 145–148
for educational programs, 113–129
for informational programs, 140–145
interface decisions, 106–107
nonsequential elements, 108–110
production script, 107–108
Setting, in narrative structure, 77
Seventh Guest, 98
Shallow branching structure, 64
Shockwave, 98, 158, 161
Shopping catalogues, interactive, 136, 149

SimCity 2000 (Maxis), 96
Simultaneity
in art, 12, 171, 179
defined, 3–4
in science, 9
Skill sets, 118–119
Slide puzzles, 94
Smart Money, 145
Soap operas, interactive, 99–100
Society and the computer, 3, 170–182
Futurists on, 171–173
Society of Mind, 40
Software
creative personnel in development of, 16
design of, defined, 16
for multimedia writers/designers, 166–169
user-testing of, 21
Spatial metaphors, 36–38
The Spot, 99, 100
Stanford University, 3, 9
Starplay Productions, 92
Stepladder branch structure, 66, 68
Story boards, 27
Story lines, in adventure games, 96, 98
Story overview, in game treatment, 103–104
Story trees, 74
Storyspace, 168
Storytelling. *See also* Narrative structure
basic elements of, 72–79
in computer adventure games, 96, 98
using hierarchical structures, 74
using linear structures, 73–74
Storyvision, 166
Strategy games, 90, 94–96
in educational multimedia, 122
Styles, interface, 78
Surrealist movement, 4–5, 172
Symmetry, in design shells, 35

T1 and T2 lines, 130, 162, 163
Table of contents, in expository structures, 81

Task
defined, 47–48
in game scripting, 103, 105–107
Teaching methods
analysis, 118–119
immersion, 118–119
Technical considerations
audio, 100, 130, 158, 161–162, 163–165
bandwidth, 98–99, 150, 161–162, 163
blue-screen process, 160–161
compression schemes, 159–161, 174
computer memory, 164–165
expanding the limits of, 165–166
format, 157–159, 163–164
graphics, 163
platforms, 157–159
playback quality, 159–160
proportions of media elements, 163
software capabilities, 165
transmission, 161–162
video, 100, 130, 159–160, 163–164
writer's literacy in, 156–157
Technology
Futurist vision of, 171–173
impact on market access, 177
social impact of, 170–174
Television, 174–176
Testing
in educational multimedia, 119, 128–129
user, 21, 27–28, 30
Tetris Max, 94, 95
Text engines, 28
Text-based games, 97, 98
Theme, in narrative structure, 78
3D JOE Corporation, 144, 145
Time Machine, 170
Tools, 47, 55–56
Top Down, 166
Tours, interactive, 141–142, 144
Training programs
advantages of, 148
analytical vs. experiential approaches to, 146

behavior or attitude-oriented training, 147–148
flowchart example for, 147–148
instructional goals in, 145–146
software for writers/designers of, 167–168
task-oriented, 146–148
Transmission technology, 161–162
Transparent interface style, 78
Treatment
for business programs, 151–153
for computer games, 102–107
content-focused, 23–25
defined, 23–24, 30
elements in, 24–25
full, example of, 107
in instructional design, 123–126
preliminary, example of, 103–107
script vs., 25
Twitch games. *See* Arcade (twitch) games
2Market, Inc., 139
2Market (CD-ROM), 139

University of California Haas School of Business interactive tour, 141, 144
University of Virginia Web Site, 130–131
User control, in interactive media, 176–177
User testing. *See also* Audience
defined, 27–28, 30

in software development, 21

Video
limitations of, 100, 130, 159–160, 163–164
playback time per script page, 164
Video disk (digital), 18, 19, 159, 164
Video games, 38
VideoDiscovery, 48
Virtual Reality Modeling Language, 57
Von Neuman, John, 9
Voyager Company, 37, 40, 74, 79, 80, 81, 82, 160

Web patterns
content arrangement and, 74–75
defined, 57
types of, 61–63
Web sites. *See also* World Wide Web
as business tool, 149, 150
as instructional tool, 130–131
software for writers/designers of, 166–168
Wells, H.G., 170
Weibel, Peter, 6, 8
Word-processing programs, 166
World Wide Web (www). *See also* Internet

authoring programs for, 167
company information on, 149
early conceptualizations of, 10
as educational tool, 129–132
format capabilities of, 18, 19, 163–164
on-line gaming and, 100
text vs. image/graphic design on, 149–150
transmission limitations of, 161–162
World-building games, 96
Writers/designers
conceptual vs. technical skills of, 167–169
educational multimedia and, 132–133
in entertainment field, 16
job title variations for, 29
as member of design team, 16–17
nature of multimedia/interactivity and, 2
production script involvement of, 27
role in technological advances, 182
software development and, 16
software tools for, 166–169
technical literacy and, 156–157
www. *See* World Wide Web

Xanadu, 10

ABOUT THE AUTHOR

DOMENIC STANSBERRY is an award-winning writer and interactive designer whose credits include *In the 1st Degree*, a multimedia courtroom drama, as well as extensive work in the fields of education, entertainment, and business for such firms as Zenger Miller, Viacom, Computer Curriculum Corporation, and Broderbund. His work has been the recipient of numerous awards and nominations, including a Cindy Gold Medal, a New Media Invision Gold Medal, an SPA Codie nomination, and a grant from the National Endowment for the Humanities. He teaches in the Multimedia Studies Department at San Francisco State University and is also the author of several works of fiction, including *Exit Paradise* and *The Last Days of Il Duce*.